BATTLE OF BRITAIN I
1940

*I*n order to establish the necessary conditions for the final conquest of England I intend to intensify air and sea warfare against the English homeland. I therefore order as follows:

1. The German Air Force is to overpower the English Air Force with all the forces at its command, in the shortest time possible. The attacks are to be directed primarily against flying units, their ground installations and their supply organisations, but also against the aircraft industry, including that manufacturing anti-aircraft equipment.

2. After achieving temporary or local air superiority the air war is to be continued against ports, in particular against stores of food, and also against stores of provisions in the interior of the country. Attacks on the south coast ports will be made on the smallest scale, in view of our own forthcoming operations.

3. On the other hand, air attacks on enemy warships and merchant ships may be reduced except where some particularly favourable target happens to present itself, where such attacks would lend additional effectiveness to those mentioned in paragraph 2, or where such attacks are necessary for the training of aircrews for further operations.

4. The intensified air warfare will be carried out in such a way that the Air Force can at any time be called upon to give adequate support to naval operations against suitable targets. It must also be ready to take part in full force in Operation Sealion.

5. I reserve to myself the right to decide on terror attacks as measures of reprisal.

6. The intensification of the air war may begin on or after 5 August. The exact time is to be decided by the Air Force after the completion of preparations and in light of the weather.

7. The Navy is authorised to begin the proposed intensified naval war at the same time.

Adolf Hitler
Directive No.17 for the Conduct of Air and Sea Warfare against England
1 August 1940

June-August 1940

Invasion?

"The British have lost the war, but they don't know it; one must give them time and they will come round."

Adolf Hitler commenting to General Alfred Jodl following the Franco-German Armistice in June 1940

June 1940 and Germany was riding the crest of a wave. Virtually everything in the campaigns against France and the Low Countries had gone according to plan. In a period of less than two months much of Western Europe had been brought under German occupation and control. All that now stood between Germany and complete domination of the whole area was Great Britain. A victorious Adolf Hitler, although preoccupied with his thoughts on the conquest of Russia, now looked for a rapid and favourable decision in the war against Britain and retained the hope that a negotiated settlement with her could be reached. However, as it became increasingly clear that no such agreement would be forthcoming, his attention began to focus on the forceful subjugation of the island kingdom, including if necessary, its invasion and occupation.

With no indication that a peaceful settlement of any kind would be reached, the OKW was now given the task of preparing for and, if necessary, achieving the successful invasion of Great Britain. For an invasion to succeed, it would have to take place within a three-month period of anticipated good weather under complete air superiority, and before the onset of the autumnal channel gales.

"The near future will show whether Britain will do the reasonable thing in the light of our victories or will try to carry on the war alone. In the latter case the war will involve Britain's destruction and may last a long time."

General Franz Halder, diary entry 22 June 1940.

RIGHT: The Bf 109 E-3 belonging to Hptm. Günther Lützow, Kommandeur of I./JG 3 at Montécouvez in May 1940. Note the early-style canopy, the six black victory bars on the rudder and the green 'Tatzelwurm' on the nose with black, red and white details. The spinner is believed to be Black-Green 70. There is no fuselage mottle and the aircraft carries a standard fuselage Balkenkreuz and Hakenkreuz.

Messerschmitt Bf 109 E-3 Hptm. Günther Lützow Gruppenkommandeur I./JG 3

The Bf 109 E-3 of the Gruppenkommandeur of I./JG 3, Hptm. Günther Lützow as seen at Montécouvez in late May. Finished in a high demarcation 02/71 scheme it carried the Gruppenkommandeur symbol ahead of the fuselage Balkenkreuz and the JG 3 'Tatzelwurm' on the cowling in the Stab colour of green. The six Abschuss bars on the rudder are believed to have been black and the spinner black-green.

I./JG 3 badge

ABOVE AND LEFT: After being hospitalised when he was shot down on 21 May, Dr. Erich Mix returned to III./JG 2 on 19 June. Once back in France, he took the opportunity to visit the wreck of his Bf 109 E-3 W.Nr. 1526 where it still lay in a field near Roye. Clearly visible in these photographs are the black-outlined Gruppe symbol and the Geschwader shield with its stylised 'R' (for 'Richthofen', the Geschwader's honour title) which had been designed for the unit by Leutnant Rotkirch.

With their record of military successes, it is not surprising that both Hitler and the OKW General Staff retained a purely continental view of carrying out such an invasion; an operation which they likened to a powerful river crossing on a broad front with the *Luftwaffe* taking the place of artillery. It was believed that the well-proven *Blitzkrieg* tactic, i.e. destruction of the opposing air force, followed by the rapid advance of the German Army with its powerful and direct air-support would also succeed against Great Britain. There was, however, one major difference – the Royal Air Force; it was the single most powerful air force yet encountered by the *Luftwaffe.* Bearing this and their recent successes against other European air forces in mind, the *Oberkommando der Luftwaffe* estimated that its complete destruction would take longer than the 12 to 48 hours taken to defeat each of the air forces fought previously. Furthermore, it was known that the British would put up a fierce and determined fight in defence of their homeland. Based on these calculations, the OKL predicted that, in the event that an invasion should take place, a period of four days would be needed to secure total air superiority in the immediate invasion area. It further anticipated that in the wake of a successful invasion, the complete destruction of the Royal Air Force could be achieved within a period of no more than three weeks. The key to the successful completion of any invasion, however, would be German supremacy in the air.

"The landing in England, prepared down to the smallest detail, could not be attempted before the British air arm was completely beaten".

General Alfred Jodl, München, November 1945

June-August 1940

ABOVE: A German soldier poses for a photograph in the cockpit of a French Caudron 714 fighter following the fall of France. The Polish national markings on the fuselage suggest that it belonged to the group of Polish volunteer pilots who fought in the Battle of France.

ABOVE: Abandoned French Caudron fighters formerly used by Polish volunteer pilots.

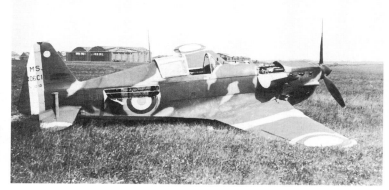

ABOVE: A damaged French Morane-Saulnier MS406C lies abandoned on an unidentified airfield following the fall of France.

ABOVE: A Dewoitine D.520 belonging to Groupement 24 GC 1/3 shot down during the battle of France. This was the first and only unit to operate the type, which became operational on 11 May 1940.

RIGHT: The remnants of France's once proud and powerful air force are collected at different locations throughout the country to be processed as scrap. The aircraft in the foreground is a Farman F 222.1.

ABOVE: An RAF Bristol Blenheim light bomber abandoned at an unidentified French location during the British withdrawal from France.

ABOVE: SS troops examine the wreck of an RAF Bleinheim shot down in northern France or Belgium during the early summer of 1940.

LEFT: An early production Hawker Hurricane Mk 1 (note the early 'pole' aerial mast) of 504 Sqn. Royal Auxiliary Air Force lies abandoned after force landing in a field somewhere in the West in the early summer of 1940. Note that souvenir hunters have been at work.

BELOW: Fairey Battles of 103 Sqn. of the RAF's Advanced Air Striking Force were stationed at Béthéniville during the early summer of 1940. This aircraft was shot down during the battle of France and has attracted the curiosity of German personnel..

ABOVE: Throughout the fighting over Dunkirk and the offensive patrols over the French and Belgian coasts, RAF Fighter Command committed its Spitfires for the first time. This photograph shows a 64 Sqn Spitfire which ended its days in a French field, providing a source of interest for local Luftwaffe personnel.

June-August 1940

'Studie Blau' (Case Blue)

In 1940 the *Luftwaffe* was undoubtedly the most powerful air force in the world but the German intelligence system was disorganised and inefficient. It was against this background that on, 1 January 1938 *Abteilung* 5, the intelligence section of the *Luftwaffe* General Staff was formed, tasked with the collection of information on foreign air forces and the preparation of target information for use in any future air war. Command of *Abteilung* 5 as Chief IC (Intelligence) was given to *Major* Joseph Schmid who, although a shrewd and ambitious man, had no foreign language skills and was not a pilot. It is perhaps significant of the value placed on intelligence by the *Luftwaffe* that the appointment required only the rank of *Major*. For the last few months of 1938 and the first half of 1939, Schmid and members of *Abteilung* 5 carried out studies on Poland, Russia and England. While the preliminary work on England was being undertaken, an order was received from the Commander in Chief of the *Luftwaffe,* Hermann Göring, demanding a high priority study of the air and industrial armaments capability of Great Britain. This subject was considered so important that a study committee was formed with Erhard Milch, Ernst Udet and Hans Jeschonnek as permanent members and Schmid as chairman. The result of this study was a full report on Britain that appeared in July 1939 under the title of '*Studie Blau*' (Case Blue). For most of the Second World War, this document would provide the basic reference material on which attacks against England were planned.

While realising that, militarily, England was an enemy to be respected, Schmid had already begun to under-estimate the efficiency and potential of the RAF. During the latter part of 1939, he devoted most of his time to formulating an offensive plan against England which would severely damage her, regardless of the outcome of the campaign against France. In late November 1939, this document was issued under the title '*Proposal for the Conduct of Air Warfare against Great Britain*'. Although far-sighted, it was probably the last study in which Schmid was able to make an objective and accurate forecast concerning Britain. While the theme of this study concentrated on the strangulation of Britain's ocean supply lines and harbour facilities he neglected to take into account two very important factors; the potential danger of the British radar chain and the lack of German maritime and torpedo bombers.

A confident and newly promoted to Reichsmarschall on 19 July, Hermann Göring delivers an address to a fighter unit somewhere in France during the mid-summer of 1940.

Oberst Josef 'Beppo' Schmid, the architect of the flawed Studie Blau.

Hermann Göring confers with his friend Ernst Udet (second from right), Josef Schmid (second from left) and an unidentified Luftwaffe officer. Both Udet and Schmid were members of the study committee set up to assess the feasibility of operations against Great Britain. It was Schmid who, as Göring's Chief of Intelligence, was largely responsible for the erroneous 'Studie Blau'.

As the time passed, Schmid concentrated on '*Studie Blau*' and the mass of information captured by Germany with the fall of France. On 16 July 1940, having compiled and studied all of the available background information, Schmid completed an overall survey on the qualities of the RAF and on which the coming offensive would be based. Aside from containing numerous misleading and inaccurate statements regarding the military and operational capabilities of the RAF, the study omitted any reference whatsoever to the closely-knit British defence system with its radar stations, operations rooms and complex HF and VHF radio network. In conclusion, Schmid stated that "... *The* Luftwaffe *is in a position to go over to decisive daylight operations owing to the inadequate air defences of the island.*"

Enigma Ultra decrypts and the Battle of Britain

In addition to an established defensive warning system, Britain had one other measure that she could employ in her defence – decoded transcripts of German signals traffic sent by the *Enigma* coding system. First coming into use during the summer of 1940, the 'Ultra decrypts' as they became known, were useful to the planning of RAF strategy but did not offer the scope of material that would become available as the war progressed. As the *Luftflotten* proceeded to follow Göring's orders, they found defending RAF fighters were nearly always there to meet them, in the right place, and, whenever possible, in significant numbers. Directed not only by the chain of RDF stations and radio communications, they were also guided by information obtained from the contents of decoded signals intercepted from the flow of *Enigma* traffic between the OKL and *Luftwaffe* units. This access to German planning strategy allowed the head of RAF Fighter Command, Air Chief Marshal Dowding and his Group Commanders, to apportion their resources accordingly and stem the *Luftwaffe* onslaught.

RIGHT: The Bawdsey Chain Home R.D.F. station which came into operation in September 1936. By February 1940, a total of twenty-nine R.D.F. stations had been completed around the coast of Britain. These lattice towers provided a primary line of sensors for watching and reporting enemy activity. There were 21 Chain Home (CH) and 30 Chain Home Low (CHL) stations plus a mobile reserve which could be used to plug gaps made by enemy attack. These towers could only detect out to sea and had no ability to look over land behind.

BELOW FROM LEFT TO RIGHT:
Air Chief Marshal Sir Hugh Dowding, A.O.C.-in-C of R.A.F. Fighter Command.

Air Vice-Marshal Sir Christopher Quintin Brand A.O.C. No.10 Group R.A.F. Fighter Command.

Air Vice-Marshal Keith Park, A.O.C. No.11 Group R.A.F. Fighter Command.

Air Vice-Marshal Trafford Leigh-Mallory, A.O.C. No.12 Group R.A.F. Fighter Command.

June-August 1940

Although the decrypts of the *Luftwaffe's* intentions clearly had an impact on the final outcome of the Battle, exactly how far they contributed to the ultimate victory of the RAF during 1940 remains an ongoing topic of discussion amongst historians to this day.

Preparation: July 1940

U ntil the end of the campaign in the West, the air war against Great Britain had been carried out on a limited scale, being confined for the most part to mine-laying, attacks against merchant and naval shipping and attacks on targets of a secondary nature. Then in June 1940, with the countries of Western Europe under German occupation, the circumstances changed. The airfields and installations of occupied Europe provided every facility to allow the full strength of the *Luftwaffe* to be strategically deployed against Britain. However, despite being in such an advantageous position, the *Luftwaffe* was faced with a formidable task, one that begged the question; "*If Britain fought on, could air power alone defeat her?*" In supporting the Army in its march across Europe, the *Luftwaffe* had played an important and decisive role in the tactic of *Blitzkrieg*. There it had been used to open each offensive by destroying the opposing air force in the air or on the ground before joining with the rapid advance of the Army to give powerful and direct air-support wherever needed. But in the mid-summer of 1940, for the first time in its history, the *Luftwaffe* would embark on a course of action that had no precedent in the history of modern warfare. Wholly independent of operations by the remainder of the *Wehrmacht*, it was to carry out an aerial offensive aimed at decisively defeating an opposing air force and forcing the capitulation of an entire nation.

> *"At this period, total air war was known only as a theoretical conception. Until then no attempts had ever been made to wage war solely by use of air power, independently of the Army or Navy, in order to break the fighting spirit of an enemy equipped with modern arms.*
> *The following were the strategic missions imposed on the Luftwaffe:*
> *a) the blockade of Britain (in conjunction with the Navy) by air attacks on shipping and ports;*
> *b) softening-up for the invasion; offensive aimed at gaining air superiority;*
> *c) forcing Britain to surrender by waging total air war against her."*

Generalfeldmarschall Hugo Sperrle, commander of Luftflotte 3.

Comment by General Adolf Galland on the role of the Luftwaffe in 1940 in a post-war appraisal of the Battle of Britain, Air Historical Branch Translation VII/121, 1953, p.11

Generalfeldmarschall Albert Kesselring, the commander of Luftflotte 2.

The necessary regrouping of *Luftwaffe* forces in preparation for the assault on England showed little change from those used in the Battle for France. *Luftflotten* 2 and 3 had simply extended their areas westwards into France with a common boundary at the mouth of the River Seine on the Channel coast. This boundary was then extended northwards through the centre of England to give each *Luftflotte* its own sphere of operations. *Luftflotte* 2 under the command of the competent and newly promoted *Generalfeldmarschall* Albert Kesselring, would operate to the east of this boundary while *Luftflotte* 3, under *Generalfeldmarschall* Hugo Sperrle, similarly recently promoted, would operate to the west. The subordinated *Fliegerkorps* of each *Luftflotte* remained unchanged but for one exception; *Fliegerkorps* *II* and *IV* were interchanged so that

Fliegerkorps IV, based in western France with units specialising in the anti-shipping role, would be better placed to operate over the shipping lanes of the Irish Sea and Western Approaches. The Norwegian based *Luftflotte* 5, under *Generaloberst* Hans-Jürgen Stumpff, would not play any major role in the early stages of the attack against Britain. However, its bombers and twin-engined fighters would provide valuable widespread anti-shipping and diversionary attacks against Northern England and Scotland which would force the RAF to keep fighter defences in the north, so weakening the aerial defences in the south.

A second change introduced with this regrouping assembled the single and twin-engined fighter units from the *Fliegerkorps* of each *Luftflotten* into two tactical fighter commands. Known as 'Jagdführer' or 'Jafüs' with *Jafü* 2 under *Luftflotte* 2 and *Jafü* 3 under *Luftflotte* 3, these commands were able to retain a measure of independence in the planning of escort duties and fighter sweeps

BELOW: The Headquarters of Jagdfliegerführer (Jafü) 2 at Le Touquet, south of Boulogne during the summer of 1940.

ABOVE: Oberst Theo Osterkamp was appointed Jagdfliegerführer (Jafü) 2 in July 1940 and in such a capacity was responsible for the tactical co-ordination of JG 3, JG 26, JG 52, JG 54 and ZG 26.

LEFT: Adolf Galland and Werner Mölders with Oberst Theo Osterkamp. Osterkamp, a veteran of the First World War, became Jagdfliegerführer (Jafü) 2 on 27 July 1940.

within the operational setting of their respective *Luftflotte.* While similar in purpose to the Fighter Groups of the RAF that they were to face in the coming battles, they lacked any means or procedure for plotting the position of enemy aircraft, nor had they any method for controlling their own fighters once airborne. Therefore, although performing the function of an operational command, aerial operations were flown without any guidance or additional direction from their ground facilities. This disadvantage would undermine the whole effort of the *Luftwaffe* Fighter Arm.

The Luftwaffe Plan of Campaign July 1940

Within the overall framework of the OKW's plans for the invasion of Great Britain, the two major tasks assigned to the *Luftwaffe* appeared straight-forward enough: neutralisation of the RAF as a fighting force and the suppression of sea-borne supplies to Britain by attacks on its ports and shipping. *Luftflotten* 2 and 3 were to achieve and maintain air superiority over southern and south-eastern England while *Luftflotte* 5 carried out diversionary attacks against Scotland and northern England to prevent some of the defending fighter squadrons from reinforcing those in the south.

Orders issued by the OKL operations staff to the three *Luftflotten* made it clear how these objectives would be met. In the first phase which would continue until the end of the first week in August, the *Luftwaffe* would attack British defences and carry out attacks against merchant and naval shipping, port

June-August 1940

A German Army band parades along Smith Street in St. Peter Port on Guernsey at the beginning of July 1940 following the occupation of the island.

facilities and selected industrial targets. This would be followed by an intensified second phase; a six week major aerial offensive designed to destroy the infrastructure and defensive capabilities of the RAF, neutralise the British coastal defences, wear down initial resistance, destroy military reserves behind the main defences and protect the build-up of invasion forces. This second phase would begin on a day given the code name *Adler Tag* – 'Eagle Day', the date of which would be determined by the first period of fine weather following the end of the initial phase. The prime objective, however, would be the neutralisation of the RAF and its ground organisation by attacking its aircraft, especially fighters, on the ground and in the air, and attacks against bases and supply installations and against its supporting industry. Once this second phase had been completed in southern England, the offensive, in keeping with the intended OKW plan, would then be extended northwards in a series of stages.

During the third week of July, the *Luftwaffe* was ordered to a state of full readiness and the final details and operational orders were worked out. The unit strength returns from the three *Luftflotten* for this week gave them a total of 2,076 serviceable aircraft of which 656 were single-engined fighters and 168 twin-engined fighters. Ranged against them was a total of 1,519 serviceable RAF aircraft of which 606 were single-engined fighters and 101 twin-engined fighters. The stage was now set for a battle that had no precedent in the history of warfare; the greatest and probably most decisive aerial battle ever fought.

1-20 July

In the broadest sense – and with some justification – it may be said that the Battle of Britain began on 30 June 1940. On that day, German forces had landed unopposed on the island of Guernsey, the largest of the four Channel Islands. Within the next 24 hours, this small group of islands, all sovereign territory of Great Britain, would be under full German control. By the end of July, an operational *Luftwaffe* landing ground had been established at the airfield on Guernsey and would be the only airfield on British soil to be used by the *Luftwaffe* during the Second World War. It would be from here that elements of JG 27 and a *Staffel* of *Major Freiherr* von Maltzahn's II./JG 53 would operate against Britain's defenders during the coming battle.

Shortly after the British evacuation at Dunkirk, Lt. Julius Meimberg and Oblt. 'Assi' Hahn of JG 2 flew from Cherbourg to Guernsey. This photograph of Meimberg with a British policeman – or 'Bobby' – and an unknown German soldier on Guernsey was taken by Hahn.

Service personnel excavating the remains of a Bf 109 E-3 of 3.(J)/LG 2 which was brought down near Sandwich, Kent during the evening of 8 July 1940. The pilot, Lt.A.Striberny baled out and was taken prisoner. Just visible on the wreckage is part of the 3.(J)/LG 2 mouse emblem which was carried on either side of the rear fuselage.

Throughout the first 20 days of July, daylight aerial activity over Britain was generally confined to bombing attacks on coastal shipping, port facilities and industrial targets, while 'freie Jagd' operations were flown by the fighters of Jafü 2 and 3. No fighters were lost in combat until the 4th, when two Bf 109s, from III./JG 27 and 4.(J)/LG 2, were claimed by P/O Smythe of 32 Sqn during a mission over the English coast. Of the two claimed, that from JG 27, although damaged, succeeded in returning to Théville, but the machine from LG 2 was seen to crash into the sea, the pilot being listed as missing. On the 5th, an aircraft from 2./JG 51 was slightly damaged in combat with Spitfires of 64 Sqn. No combat losses occurred on the 6th but during the evening of the 7th a 'Freiejagd' by JG 27 resulted in the downing of three Spitfires of 64 Sqn. Although two victories were claimed by RAF pilots, the only Bf 109 E listed as damaged that day, was an aircraft of III./JG 27 which was severely damaged in a take-off accident at Théville.

On the 8th, increased activity over Channel convoys resulted in the loss of three fighters including the first Bf 109 E to come down on British soil. At 15.45 hrs this aircraft, 'White 4', an E-3 of 4./JG 51 flown by Lt. Johann Böhm, force landed at Bladbean Hill, Elham, Kent after being damaged by a Spitfire of 74 Sqn. This was followed at 19.30 hrs by the second Bf 109 to crash in Britain, this time an aircraft from 3.(J)/LG 2, which was shot down by 54 Sqn and crashed near Sandwich, Kent. The pilot, Lt. Albert Striberny, was captured after baling out. An aircraft from III./JG 51 was also lost in action on this date with a fourth from II./JG 51 being damaged in combat with 610 Sqn. The 9th again saw increased fighter activity over the Thames Estuary and Channel but, despite RAF claims for two Bf 109s shot down, only one, from II./JG 51, was recorded as being lost.

> *"...he saw me almost immediately and rolled out of his turn towards me so that a head-on attack became inevitable. Using both hands on the control column to steady the aircraft and thus keep my aim steady, I peered through the reflector sight at the rapidly closing enemy aircraft. We opened fire together, and immediately a hail of lead thudded into my Spitfire. One moment the Messerschmitt was a clearly defined shape, its wingspan nicely enclosed within the circle of my reflector sight, and the next it was on top of me, a terrifying blur which blotted out the sky ahead. Then we hit."*
>
> F/Lt. Al Deere, 54 Sqn, commenting on his head-on collision with a
> Bf 109 of II./JG 51 during a dogfight with the fighter escort for
> a He 59 of Seenotflugkommando 1 during the evening of 9 July 1940

On 9 July 1940, this Heinkel He 59 float plane (D-ASUO), was forced down on the Goodwin Sands by P/O J.L. Allen of 54 Sqn. It was later towed to the beach at Deal by the Walmer lifeboat.

June-August 1940

The monument at Cap Gris Nez near Wissant was often used as a vantage point by high-ranking German officials to observe the British Isles. Behind the monument in this view can be seen a FuG 401 Freya radar station.

Luftwaffe Fighter Disposition during the Battle of Britain – Pas De Calais

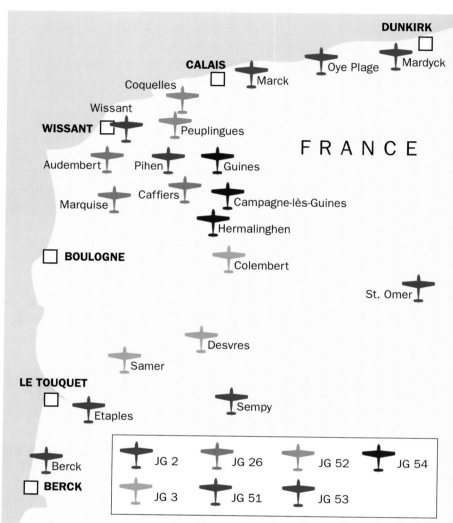

DUNKIRK

Oye Plage Mardyck

CALAIS Marck

Coquelles

Wissant

WISSANT Peuplingues

FRANCE

Audembert Pihen Guines

Caffiers Campagne-lès-Guines

Marquise

Hermalinghen

BOULOGNE Colembert

St. Omer

Desvres

Samer

LE TOUQUET Sempy

Etaples

Berck

BERCK

JG 2		JG 26		JG 52		JG 54
JG 3		JG 51		JG 53		

Luftwaffe Fighter Disposition during the Battle of Britain – Normandy, Brittany and Channel Islands

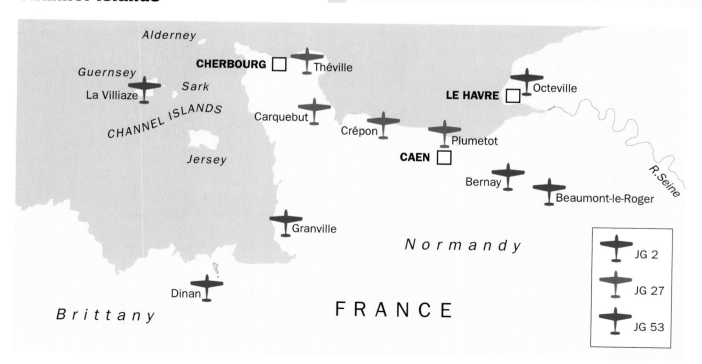

Alderney

Guernsey CHERBOURG Théville

La Villiaze Sark Octeville

CHANNEL ISLANDS Carquebut LE HAVRE

Crépon Plumetot

Jersey CAEN

Bernay R. Seine

Beaumont-le-Roger

Granville

Normandy

Dinan

Brittany FRANCE

JG 2
JG 27
JG 53

June-August 1940

On 10 July, the major fighter activity of the day took place over the west-bound convoy 'Bread' in the English Channel. There were no German fighter losses during the morning's fighting but a JG 51 aircraft was claimed as slightly damaged by a Spitfire from 74 Sqn. In the afternoon battles above the convoy, one Bf 109 from 5./JG 51 was lost while a further two from 7./JG 51 were damaged severely enough to result in forced landings in France, both pilots escaping serious injury.

ABOVE: An early-style life jacket is stored ready for use in the cockpit of a Bf 109 E somewhere in France in the Summer of 1940. Note how the 8 mm armour plate - as fitted to the older style hood - adversely affected vision to the rear.

"The whole cockpit stank of burnt insulation but I managed to stretch my glide to the coast, then made a belly-landing close to Cherbourg. As I jumped out the machine was on fire, and within seconds ammunition and fuel went up with a bang!"

Ofw. Arthur Dau, 7./JG 51, commenting on what happened after being hit by fire from a Hurricane flown by Sgt. A.G. Page, 56 Sqn during combat over the convoy 'Bread' 10 July 1940

"Suddenly the sky was full of British fighters. Today we were going to be in for a tough time."

Hptm. Hannes Trautloft, III./JG 51, commenting on the action over the British convoy 'Bread' 10 July 1940

LEFT: Personnel of the Stabsschwarm of III./JG51 who were involved in the battle with the Defiants of 141 Sqn on 19 July 1940. They are left to right: Oblt. Kahn, Adjutant, Oblt.Pichon-Kalau von Hofe, Technical Officer, Oblt. Wehnelt and Hptm. Hannes Trautloft the Gruppen-kommandeur.

On the 11th, 12th and 13th, attacks on Channel convoys continued resulting in one Bf 109 of 9./JG 51 being shot down near Dover on the evening of the 13th. On the 14th, battles again developed over Channel convoys resulting in the loss of one aircraft from 8./JG 3 and a second from the same *Staffel* returning to France severely damaged. Although skirmishes continued for the next four days, operations were hampered by bad weather and no further fighters were lost in combat until the 18th when a Bf 109 of II./JG 2 was lost to unspecified reasons during an operational sortie. On the 19th, the day on which Hitler would make his 'Last Appeal to Reason' speech, improving weather resulted in increased fighter activity off Dover. In the early afternoon, the Defiants of 141 Sqn were badly mauled by fighters of III./JG 51 and II./JG 2 for the cost of one aircraft of 9./JG 51 severely damaged. In later actions over Folkestone and Selsey Bill, three more fighters from 9./JG 51 and III./JG 27 were badly damaged with two of the pilots being wounded. Improving weather over the Channel on the 20th saw an early afternoon attack on Dover which claimed two aircraft from 3./JG 27. The *Geschwader* suffered a third loss during the late afternoon when the *Gruppe Kommandeur, Major* Riegel, was shot down off the island of Sark by two Hurricanes of 501 Sqn. Later, in an early evening battle that developed over the convoy 'Bosom,' two more aircraft, both from JG 51, were lost to RAF fighters.

BELOW: Photographed in July 1940, this view shows the emblem carried on the starboard escape hatch of Hurricane P3878, YB-W of the Debden-based 17 Sqn. On 24 September while being flown by P/O H.A.C. Bird-Wilson it was shot down over the Thames Estuary by Adolf Galland to become his 40th victory. P/O Bird-Wilson although burned, was able to parachute to safety and was admitted to the Royal Naval Hospital at Chatham.

ABOVE:
Pilots of 32 Sqn photographed at Hawkinge at the end of July 1940. From left to right they are:
P/O R.F.Smythe,
P/O K.R.Gillman,
P/O P.M.Gardner,
P/O J.E.Proctor,
F/Lt. P.M.Brothers,
P/O D.H.Grice and
P/O A.F.Eckford.
All would survive the war except Keith Gillman, MIA on 25 August 1940.

21-31 July

In contrast with the previous day's activity, the 21st was relatively quiet until mid-afternoon when the west-bound convoy 'Peewit' came under heavy attack from elements of KG 3 south of the Needles. During the ensuing battle between the fighter escort and the defending fighters, one Bf 109 from 7./JG 27 was lost when it collided with a 43 Sqn Hurricane flown by P/O DeMancha. Both pilots were killed in the collision. A second loss also occurred this day when a Bf 109 from III./JG 77 was lost for unspecified reasons while on an operational sortie over the North Sea. There was little fighter activity over the Channel on the 22nd and 23rd but on the 24th, Channel convoys were again the focus of *Luftwaffe* attention. An intense fighter battle developed over a convoy in the Straits of Dover which cost JG 26 three aircraft. *Oblt.* Werner Bartels, the *Geschwader Technische Offizier (TO)* force-landed near Margate and was taken prisoner, *Lt.* Josef Schauff was killed when his aircraft crashed in Margate and his parachute failed to open and *Hptm.* Erich Noack, *Kommandeur* of the second *Gruppe,* was killed when his Bf 109 crashed while attempting to land at Marquise-East.

Four aircraft from JG 52 were also lost in this action, one of these being *Hptm.* Wolf-Heinrich von Houwald the *Kommandeur* of III./JG 52. On the 25th, another ferocious battle took place over a convoy passing through the Dover Straits that would last through most of the afternoon. This action saw the loss of seven fighters; one from 9./JG 26, one from III./JG 27 and four from III./JG 52. The seventh loss recorded for the day was a severely damaged E-1 from 5./JG 51 which was written off after returning to St. Inglevert. Deteriorating weather over the Channel on the 26th and 27th limited attacks

LEFT: Although the precise date is undetermined, this photograph shows a line of replacement Bf 109 Es for the Caffiers-based 8./JG 26. As the third Gruppe of JG 26 retained the high demarcation Blue 65 finish throughout 1940, the fuselage Balkenkreuz and numerals were reapplied in smaller format to help conceal the aircraft at high altitude.

in any strength against convoys and other shipping. On the 27th attacks were carried out against both Dover and the convoy '*Bacon*' but the only fighter lost in combat during these two days was an aircraft from 2./JG 27, shot down south of Portland by a Hurricane of 238 Sqn at noon on the 26th.

> *"We were no longer in doubt that the RAF would prove a formidable opponent"*
>
> Adolf Galland, III./JG 26, commenting on his
> first combat over the English Channel, 24 July 1940

Despite better weather over the Channel on Sunday 28th, no concentrated attacks on shipping developed until early afternoon when an incoming raid was detected heading for Dover. In the battle that followed one machine from 2./JG 51 was lost and two others badly damaged, of which one was flown by the *Geschwaderkommodore, Major* Werner Mölders. Also involved in the fighting this day was the highly respected and popular South African from 74 Sqn, F/Lt. A.G. 'Sailor' Malan. While some sources have credited Malan with damaging the Bf 109 flown by Mölders, others credit the action to Fl./Lt.J.T.Webster of 41 Sqn. Although a detailed study of combat reports for this engagement suggest that Malan may have been responsible, it is far more conceivable that the damage was actually inflicted by Webster.

ABOVE: A Bf 109 pulls away after shooting down one of the Dover barrage balloons during the afternoon of 31 July 1940.

> *"North of Dover we met some low-flying Spitfires. I shot down a Spitfire in flames. But now I found myself in the middle of a clump of Englishmen and they were very angry with me. They all rushed at me and that was my good luck. As they all tried to earn cheap laurels at the expense of one German, they got in each other's way. Well, I managed to manoeuvre among them and made them even more confused. Nevertheless, I couldn't avoid being hit. Bullets bespattered my aircraft. The radiator and fuel tank were shot up badly and I had to make a getaway as quickly as possible. Luckily my engine held out to the French coast, then it began to misfire. When I wanted to land, the undercarriage wouldn't work. There was nothing to do but land without it. I made a smooth belly landing."*
>
> Major Werner Mölders, Stab/JG 51, commenting on his
> first combat over the Channel, 28 July 1940

BELOW: Reflecting a sight that would become common to both sides during the summer air battles, two German airmen of 7./KG 55 paddle their dinghy through the unforgiving waters of the English Channel off Shoreham. Their He 111 had firstly been engaged by 1 Sqn and eventually shot down by aircraft from 145 Sqn.

With more fine weather early on the morning of the 29th, Dover harbour and two Channel convoys were targeted by the *Luftwaffe*. A heavy raid in the early morning was directed at Dover but was driven off by the anti-aircraft and fighter defences and afternoon attacks carried out against the two convoys

caused little damage. Activity over the Channel on this day resulted in four fighters being severely damaged in combat. Of these four, one from I./JG 51 crashed at Wissant and one from 6./JG 51 crashed outside Calais with both pilots being killed. The two other aircraft from II./JG 27 and 4./JG 51 force-landed without injury to either pilot. With low cloud and light rain covering most of Britain on the 30th, air activity was greatly reduced and no Bf 109s were lost. Although the weather began to improve on the 31st, hazy conditions frustrated operations. Later that afternoon two *Staffeln* of Bf 109s from JG 2 shooting up barrage balloons in the Dover area were intercepted by Spitfires of 74 Sqn. None were shot down but one Bf 109 was damaged and force-landed at Fécamp with a seized engine.

June-August 1940

1-12 August

With a day of low cloud and mist heralding the beginning of a new month, the *Luftwaffe* paid little attention to the convoys around Britain and it was not until the mid-afternoon of 1 August that the convoys 'Agent' and 'Arena' provoked any reaction. During the day no fighters were lost in action but three aircraft of II./JG 27 were damaged to varying degrees in an RAF Bomber Command attack on

Leeuwarden airfield in northern Holland. Continuing poor weather on the 2nd, 3rd and 4th again limited *Luftwaffe* offensive operations and there were no fighter combat losses. Fine weather early on the morning of the 5th saw a furious engagement develop over the Kent coast between Spitfires of 64 Sqn and aircraft from JG 54, which resulted in two Bf 109s returning to France damaged. Later in the day, a battle above a convoy in the Straits of Dover saw one aircraft of JG 51 shot down and another returning to its base damaged. On the 6th and 7th aerial activity was again limited as far as fighter actions were concerned. The only operational fighter casualty during these two days was an aircraft from JG 3 which was damaged when it force-landed after an operational sortie on the 6th.

An undated photo of Werner Mölders (right, in flying jacket and gloves), in conversation with other members of JG 51.

"Why marry now when there is only England left?
Marry later to celebrate the victory".

Major Werner Mölders, Stab/JG 51, replying to
a request from one of his pilots seeking leave to marry, 7 August 1940

On the morning of the 8th and marking the start of a distinct new phase of attacks, the west-bound convoy 'Peewit' was subjected to a series of attacks more intensive than any made against convoys during the preceding month. During the course of the day, three furious air battles took place over and around the convoy. At the end of the day's fighting, nine fighters had been lost with a further eight damaged to varying degrees. Of the units taking part, II./JG 27 suffered the most casualties with four aircraft lost and two damaged, one of these losses being the *Gruppenkommandeur*, *Hptm*. Werner Andres who survived ditching his aircraft and was later rescued from the Channel by the *Seenotdienst*.

"The enemy fighters, which were painted silver, were half-rolling and
diving and zooming in climbing turns. I fired two five-second bursts at
one and saw it dive into the sea. Then I followed another up a
zoom and got him as he stalled"

S/Ldr. John Peel, 145 Sqn commenting on
the battle above the convoy 'Peewit' 8 August 1940

S/LDR. JOHN PEEL (BRITISH), 145 SQN. RAF

John Peel was born on 17 October 1911 and entered the RAF College at Cranwell as a Flight Cadet in September 1930. He graduated in July 1932 and joined 19 Sqn at Duxford shortly after. In January 1934, he was posted to 801 (Fleet Fighter) Sqn, flying alternatively from the airfield at Upavon or from the carrier HMS *Furious*. He joined 601 Sqn, Royal Auxiliary Air Force, in September 1935 where he spent time as a flying instructor before moving on to a staff position at Cranwell in July 1936. He returned to 601 Sqn in September 1937.

In July 1940, while serving on the staff of the Postings Section of the Air Ministry, Peel was given command of 145 Sqn. On the 7th of that month, he shared in the destruction of a Do 17. On the 11th, Peel is believed to have shot down either a Bf 110 or a Do 17, but was himself shot down in this action. He ditched in the Channel off Selsey Bill and was rescued by the Selsey lifeboat.

On the 17th, he claimed a Ju 88 as damaged and shared in the destruction of another Do17 on the 19th. Again, on the 29th he shared in the destruction of a Ju 88 and on 8 August claimed two Ju 87s and one Bf 109 destroyed. He was awarded the DFC on 30 August. On 16 December he left 145 Sqn for a brief period before returning again as its commanding officer in November 1941. He survived the war to retire from the RAF on 20 January 1948 with the rank of Group Captain.

June-August 1940

F/Lt. Adolf Gysbert 'Sailor' Malan (South African), 74 Sqn. RAF

Born in Wellington, South Africa on 3 October 1910, Adolph Gysbert Malan became a cadet on the training ship *General Botha* in February 1924 and joined the Union Castle Steamship Line in 1927. In 1935, he applied for and received a short service commission in the RAF and began his flying training at No.2 E & RFTS Filton on 6 January 1936. From Filton, he went on to No.3 FTS at Grantham, was posted to 74(F)Sqn at Hornchurch on 20 December 1936, and was promoted to Flight Commander in late 1937.

Near Dunkirk on 21 May 1940, he destroyed a Ju 88, claimed the probable destruction of a He 111 and damaged a second Ju 88. On the 24th, he shared in a victory over a Do 17 and claimed a He 111 destroyed. On the 27th of that month, he claimed one Bf 109 destroyed, shared a probable Do 17 and damaged two others. He was awarded the DFC on 11 June.

During the night of the 18/19 of June, he destroyed two He 111s and on 12 July, he shared in the destruction of a third. On the 19th, he claimed the probable destruction of a Bf 109. On the 25th, he claimed another Bf 109 as damaged and on the 28th, destroyed one Bf 109, claiming a second as damaged. On 8 August, he took command of 74 Sqn. On the 11th, he claimed two Bf 109s as destroyed and one damaged and on the 13th claimed the destruction of one, and possibly two Do 17s. On 11 September, he destroyed a Ju 88 and damaged a second then, on 17 October, claimed a Bf 109 as a probable. On the 22nd, he destroyed a Bf 109 which was followed by the destruction of another on the 23rd. He destroyed one Bf 109 and shared in the destruction of another on the 27th and claimed a further Bf 109 on 2 December. He was awarded the DSO on 24 December.

On 21 July 1941, Malan was awarded a Bar to his DSO and in October of that year, was sent on a tour of the USA with five other pilots to lecture and liaise with the US Army Air Corps. He survived the war to retire with the rank of Group Captain and returned with his family to South Africa in 1946. In addition to his British awards, he was awarded the French and Belgian *Croix de Guerre*, the French *Legion d'Honneur* and the Czech Military Cross. He died in South Africa on 17 September 1963.

Activity over the Channel on the 9th and 10th was slight and no fighters were lost on operations, although a Bf 109 of I./JG 53 was written-off on the 9th when it hit a *Flak* emplacement while making an emergency landing on Guernsey due to engine failure at the end of an operational sortie. Improved weather on the 11th again saw the *Jagdwaffe* out in strength undertaking 'Freiejagd' over the Kent and Sussex coasts and escorting attacks carried out on Portland, Weymouth and Channel convoys. Intensive fighting developed during the early morning and continued throughout the day until deteriorating weather in the afternoon brought most aerial activity to an end. The aircraft lost in the actions of the 11th almost doubled those of the 8th, totalling 14 fighters lost and two damaged. The losses to JG 2 alone almost equalled those incurred on the 8th and included the *Gruppe Adjutant* of III./JG 2, *Oblt*. Adolf Steidle and the *Staffelkapitän* of 6./JG 2 *Oblt*. Edgar Rempel.

With *Adler Tag* set for the following day, the morning of the 12th dawned bright and clear. This day would witness the first attacks to be carried out against RAF airfields and coastal RDF stations. Set in a series of stages, the attacks moved back and forth along the south coast throughout the day. The airfields of Hawkinge, Lympne and Manston came under heavy attacks and Ventnor RDF station, hit heavily at around midday, was put out of action. Pressure was also maintained in the attacks against coastal shipping and harbours with the convoys 'Arena', 'Agent', 'Snail' and 'Cable' receiving particular attention. Despite the ferocity of the day's fighting, fighter losses were lower than those for the 11th. In all a total of nine Bf 109s failed to return from the day's actions including that flown by the *Gruppenkommandeur* of III./JG 53, *Hptm*. Harro Harder. A further five aircraft received various levels of combat damage, one of which, from I./JG 2, was subsequently written off.

The attacks carried out against the airfields and RDF stations during the 12th were severe and gave a foretaste of what lay ahead. Although the airfields of Hawkinge, Lympne and Manston were serviceable the next day, it was a full three days before, Ventnor RDF would function again, leaving a vital breach in the British warning system. Fortunately this loss was disguised from German signals intelligence by having the Ventnor signals transmitted by another station until repairs were completed at the site.

June-August 1940

ABOVE: The port side of the Bf 109 E of Oberfeldwebel Werner Machold of 7./JG 2 showing his eight victories.

LEFT: A posed propaganda photograph of Oberfeldwebel Machold pretending to paint his ninth victory on the rudder of his Bf 109 E at Cambrai on 27 May 1940.

RIGHT: German Army personnel pose for a photograph on an abandoned French MS 406 at an unidentified location following the fall of France.

ABOVE: Uffz. Rudolf Rothenfelder and Fw. Wieser of III./JG 2 take a rest on the wreckage of a Bloch at Couvron on 10 June 1940.

LEFT AND ABOVE: The wreckage of Bf 109 E Yellow or Brown '2', probably of I./JG 52 lies in a French field in June 1940 following the air fighting over Dunkirk.

June-August 1940

LEFT: Taken on 13 May 1940, this photograph shows Oblt. Hubertus Freiherr von Holtey, Staffelkapitän of 5./JG 26 standing beside the starboard wing of a Bf 109 E which carries two emblems on its starboard side. Ahead of the windscreen is the white JG 26 'Schlageter' shield with its script black 'S' while beneath the canopy is the 5.Staffel emblem of the 'Hans Huckebein' cartoon raven character which is black with white eyes and a white or yellow beak.

RIGHT: Taken at the end of the Battle for France, this photograph shows 9./JG 2 dispersed at a rudimentary landing ground near Signy-le-Petit surrounded by their ground support vehicles. Note how on the aircraft in the foreground 'Yellow 5', W.Nr.1146, foliage and netting has been used to break up the outline of the aircraft. This aircraft survived the Battle of Britain but was lost on 17 April 1941 while on the strength of 1./JG 1 when it crashed in the North Sea, killing the pilot, Ofhr. Friedhelm Gottschalk.

Messerschmitt Bf 109 E–3 of 9./JG 2

Bf 109 E-3 'Yellow 5' of 9./JG 2 as seen at a landing ground near Signy-le-Petit at the end of the Battle of France in a high demarcation 02/71 scheme. Although listed as an E-3 it retains the earlier style of canopy and appears to be carrying its Werknummer on the rudder instead of the more commonly seen location on the fin. This aircraft survived the summer battles only to be lost in an accident on 17 April 1941 while on the strength of 1./JG 1.

**JG 2 Richthofen
Geshcwader badge**

III./JG 51
'Axt von Niederrhein'
emblem

Messerschmitt Bf 109 E-3 Oblt. Werner Pichon-Kalau von Hofe, Technical Officer III./JG 51
Oblt. Werner Pichon-Kalau von Hofe's Bf 109 E-3 as it appeared during the mid-summer of 1940. Finished in a high demarcation 70/71 camouflage scheme it carried the III.Gruppe 'Axt von Niederrhein' beneath the cockpit ledge.

8./JG 51
'Black Cat'
emblem

ABOVE AND RIGHT: Bf 109 Es of III./JG 51 on an unidentified French airfield in the early summer of 1940. The aircraft in the middle foreground is that of the Technical Officer, Oblt. Werner Pichon-Kalau von Hofe. The black cat emblem on the cowling of the aircraft in the foreground is that of 8. Staffel which had been 2./JG 20 before it became a part of JG 51. Both aircraft are finished in the high demarcation 71/02/65 scheme.

June-August 1940

LEFT: Armourers install an ammunition drum in the wing of a Bf 109 E-3. The wing cannon are 20 mm MG FFs.

BELOW: A mechanic carries out maintenance to the Daimler-Benz DB 601 engine on a Bf 109 E-3 somewhere in France during 1940.

BELOW: Ground crew undertake synchronisation adjustments to the MG 17s mounted above the Daimler-Benz 601 engine this Bf 109 E-1, red 10 outlined in white..

"I could see the entire British island..."

Konrad Jäckel, JG 26

I was born on 12 July 1917 in Limbach, Saxony. At the age of 12, I became a member of the Hitler Youth in Limbach. I joined the *Flieger Jugend*. Since I had skills from working with my hands, I built models. As a student, I continued building models for which I was awarded first prize on several occasions. I attended several courses in model and glider construction. I volunteered for the *Luftwaffe* and enlisted for four and a half years, thus avoiding the *Arbeitsdienst*. I reported to Dresden for my examinations, and six months later, I reported for duty in Breslau, Silesia. I had no idea I would become a pilot since I assumed that I would be assigned as a mechanic. However, since I apparently had the qualities required to become a pilot, I was assigned to flight training.

In Breslau, I qualified for the 'A' flight certificate. Then I received infantry training, followed by the 'B' flight certificate at Schleissheim, near München. At Neuruppin I received my 'C' flight certificate when I flew the Ju 52, He 111, and other aircraft. Then I was asked as to which branch of the *Luftwaffe* I wanted to be assigned and I requested fighters. I was assigned to a fighter training school where I met pilots who had fought against the Communists in Spain. One of our flight instructors was a lieutenant who had scored seven kills there.

When 4./JG 26 was formed, *Oberleutnant* Edu Neumann was the *Staffelkapitän*. I was assigned to another unit but Neumann requested my return. The unit was later redesignated 8./JG 26 which was nicknamed the '*Adamson Staffel*'. After Neumann had returned from Spain, he introduced the *Adamson* emblem. Our *Gruppe* commander was *Hptm*. Werner Palm who was relieved of his command because he had no fighting spirit.

My first kill was on 28 May 1940 over Ostende, Belgium. We were engaged in an air battle when a Hurricane appeared behind *Hptm*. Müncheberg. I shot down the Englishman who would otherwise have shot down Müncheberg.

I flew 115 combat sorties of which between 75 and 80 had contact with the enemy. The British didn't always intercept our flights. Only when there were bombers in the air would they rise to fight us. I was Gustav Sprick's wingman. I was responsible for securing the area behind him and also to confirm his kills. This meant that I had a dual responsibility. Ebeling took command of a *Staffel* which was re-equipped to carry bombs. He had 20 kills at the time. We escorted his *Jabo Staffel* which the English would intercept.

Our unit was stationed north of Cap Gris Nez. We were billeted in a monastery. On our first mission to England, I had this odd feeling while flying over the waters of the English Channel. I already had experience flying over water when I used to land at Swinemünde where I made the approach flight to the airfield over the sea.

On one mission to London, I was separated from the *Staffel* and for security, I climbed to 11,000 metres. The weather was beautiful and there were few clouds. I could see the entire British island and the silver reflections from the rivers. With such marvellous scenery, I still had to watch out for enemy fighters as I flew home.

A Bf 109 E-3 possibly of JG 2 gets airborne from a landing strip in France during the summer of 1940. Although finished in what appears to be an 02/71 upper scheme with mottled fuselage sides, the swastika has been retained in the earlier position.

ABOVE: Taken in July 1940, this photograph shows the Bf 109 E of Hptm.Willi Meyerweissflog, (W.Nr. 5377) of Stab/JG 53 being refuelled from 45 gallon drums carried on the back of a Commer lorry which has been pressed into temporary airfield service at La Villiaze on the island of Guernsey.

ABOVE AND RIGHT: The Bf 109 E-1 of Lt. Johann Böhm of 4./JG 51 at Bladbean Hill, Kent seen the day after it was brought down. To right is a close-up view of the rear fuselage and the 4.Staffel emblem and at left, now raised up on its undercarriage to form a photographic background for these members of the R.A.F. recovery team.

Emblem of 4./JG 51

Messerschmitt Bf 109 E-1 Lt. Johann Böhm 4./JG 51

This was the first Bf 109 E-1 to come down intact in the British Isles when on 8 July, after being damaged by a Spitfire of 74 Sqn flown by Sgt.E.A.Mould, Lt. Böhm was forced to put his crippled aircraft down at Bladbean Hill at Elham in Kent. Although not very clear in photographs, the upper surface camouflage of this aircraft appears to be 70/71 with the fuselage sides carrying a heavily applied mottle believed to be a combination of one of the upper-surface greens and 02. The white number '4' is outlined in red and the 'Weeping Bird' emblem is closer to the Balkenkreuz than usually seen.

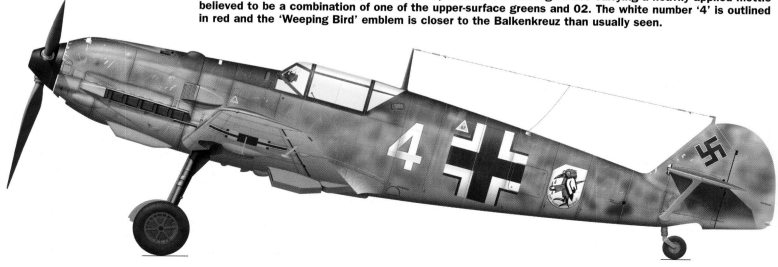

Messerschmitt Bf 109 E-3 of I./JG 27
Bf 109 E-3 'Yellow 11' of 3./JG 27 understood to have been based at Plumetot during July 1940. Finished in a high demarcation 02/71 scheme the rudder of the aircraft is a very dark colour suggesting a replacement item which still carries a coat of fabric primer. Just ahead of the aircraft number can be seen a pair of white scissors which is understood to represent a play on the name of Lt.Ulrich Scherer, a 3.Staffel Schwarmführer who was killed in action on 20 July. On the cowling is the JG 27 'Afrika' emblem consisting of the heads of a Negress and a Leopard superimposed on a map of Africa. The origins of this emblem can be connected to either Hauptmann Helmut Riegel or Oberleutnant Gerd Framm both of whom had family connections with the former German colonies in Africa.

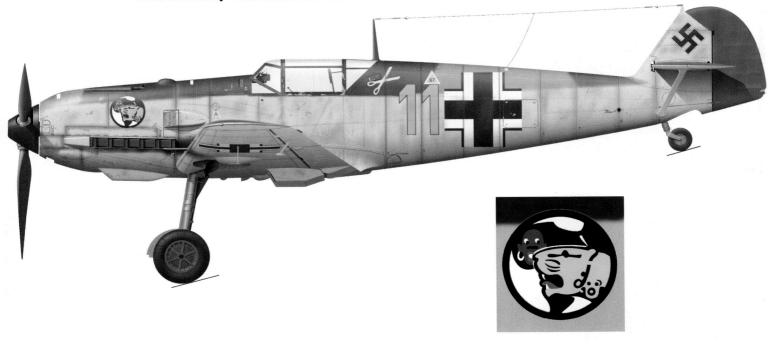

**I./JG 27
Gruppe badge**

BELOW: Flown by Oblt.Werner Schuller, Bf 109 E-3 W.Nr. 3225, 'Yellow 11' of 3./JG 27 gets airborne from an airstrip in Normandy, probably Plumetot, sometime during July 1940. The dark rudder is believed to be a replacement still in primer finish (RLM 02). Also visible are the white scissors immediately behind the cockpit glazing.

June-August 1940

Messerschmitt Bf 109 E-4 of 3./JG 2

Leutnant Franz Fiby's Bf 109 E-4 'Yellow 14' of 3./JG 2 with an upper surface scheme that is believed to have been 02/71 with heavy mottling of 71 on the fuselage sides and tail. Note how the broad white areas of the Balkenkreuz have had their visibility reduced by being overpainted with the fuselage mottling, making the cross appear reminiscent of the earlier style of Balkenkreuz although the thin outer black borders are still faintly visible. Also carried are the JG 2 'Richthofen' shield beneath the windscreen and on the cowling, the blue and yellow 3.Staffel 'Horrido' pennant which would later be adopted by Helmut Wick as his personal emblem.

**3./JG 2 'HORRIDO'
Staffel emblem**

ABOVE: Franz Fiby's Bf 109 E-4 of 3./JG 2, 'Yellow 14', over France in July 1940, wearing a dark, heavily stippled camouflage on the fuselage sides. Close examination of the fuselage cross shows that its non-standard appearance and thin black borders have been created by the overpainting of the usually seen white segments of the cross.

LEFT AND OPPOSITE: The Bf 109 E-1 W.Nr. 6296F of Oblt. Werner Bartels, Technical Officer of III./JG 26 lies in a field at Northdown, Kent close to the Broadstairs-Margate railway line following a forced-landing on 24 July after being attacked by a Spitfire. The 'F' suffix to the W.Nr. stands for Flugklar, (literal translation 'Flight ready') denoting a repaired or rebuilt aircraft which has been cleared to return to front-line duties. The photos clearly show that the new paintwork was applied around many of the original stencil marking. It is unusual to find this number of photos of a shot machine, showing both sides and the various places the machine was put on show before being scrapped.

Messerschmitt Bf 109 E-1 W.Nr.6296F Oblt.Werner Bartels Technical Officer III./JG 26
A rebuilt or repaired machine as identified by the 'F' suffix to the W.Nr. the aircraft is fitted with the heavier framed canopy and windscreen more usually seen on E-3 and E-4 variants. The upper surface camouflage is understood to have been 02/71 which, like the heavier canopy, may well have been applied when the aircraft was repaired.

ABOVE AND ABOVE RIGHT: In these two photographs, Oblt.Werner Bartels' Bf 109 E-1 has now been raised and had its undercarriage extended in preparation for removal from the site of its forced-landing.

RIGHT AND BELOW: Oblt.Bartels' Bf 109 E-1 is put on public display at Croydon following its recovery from the field at Northdown.

BELOW: Early 1941 and Oblt. Bartels' Bf 109, now in much worse condition than when it was captured, is still being exhibited around Britain.

ABOVE: Another view of Werner Bartel's machine at Croydon, this time being used in a Red Cross publicity photograph. During and after the Battle of Britain, many captured Luftwaffe aircraft were exhibited for either propaganda purposes or to serve as backdrops to fund-raising events for war bonds or the well-publicised 'Spitfire Fund'.

"I closed to within 20-30 metres and fired with both machine guns and both cannon. The upper gunner really had some nerves..."

Hans Schmoller-Haldy, JG 54

At the time of the armistice, we were near Paris. That evening I was awoken by the officer-of-the-day and was told that the following morning we would be flying to Eindhoven in the Netherlands. The *Staffel* took off at 09.00 hours and we landed at Eindhoven at 11.00 hours. We made an intermediate landing to refuel. The Battle of Britain was about to begin.

On 26 June 1940, I was wounded in the leg. I had taken off from Eindhoven and was flying in formation toward Germany. A British Blenheim was reported to be flying over Holland. I was 500 metres above him near Amsterdam. I reported the British aircraft to the others in the *Schwarm*. I banked and dived and came in immediately behind him just like I had practiced during training. The gunner started firing at me. I rapidly closed to within 20-30 metres and fired at him with both machine guns and both cannon. The upper gunner really had some nerves and continued firing at me. As long as I was directly behind him, he could not fire through his fin and could only fire at my wings. I was closing too rapidly and I would have rammed him, so I had to turn away. I could see his tracers and the gunner had a good target when I showed him my underside. My engine was shot up and I felt a sharp pain in my leg. Fortunately, the aircraft did not burn. I pulled out at low level. I reduced the throttle. I was losing blood and I reported on the radio that I was wounded. I landed at Eindhoven and was immediately taken to the hospital there. It is amazing that the Blenheim was able to continue flying for another 200-300 kilometres.

I lay in hospital near Eindhoven for two weeks. I returned to flying in August.

I had to fly the first missions over England with my leg wrapped in bandages. During a six-week period, all of the pilots in the *Staffel* except my deputy and I, were lost. The last sorties were flown by only two or three pilots. The British were now so strong that they could meet us half way across the Channel. We had started the French campaign with 12 pilots but now I had only *Lt.* Kitzinger and *Fw.* Knipscher. On the last mission, Knipscher was shot down. Kitzinger and I transferred to Jever. The base commander, an *Oberstleutnant*, greeted me and told me to bring in my *Staffel*. *Lt.* Kitzinger was standing in the doorway and I pointed to him and said: "This *is* my *Staffel*!"

ABOVE: The Château de Colombert used as the headquarters for I./JG 3.

LEFT: The Château at le Colombier near Audembert, which was used by Adolf Galland's III./JG 26.

BELOW: The Château du Désert near Desvres, used as a headquarters by Stab III./JG 3

June-August 1940

LEFT: For a short time following the conclusion of the campaign in the West, the name of Oblt. Paul Gutbrod was carried by aircraft of II./JG 52. Gutbrod, who was a member of the Stab II./JG 52, was killed on 1 June 1940 and was honoured in recognition of the fact that he opened the Gruppe's victory tally. In common with several other Bf 109s, this machine carries a diagonal yellow line from the rear of the cockpit to the trailing edge of the wing. The significance of this is not known.

RIGHT AND BELOW: Damaged Bf 109s – 'White 4' of I./JG 27 and 'Yellow 6' of I./JG 52 – stand parked next to a damaged French Mureaux 115, during the early stages of the Battle of Britain.

June-August 1940

**7./JG 2
Staffel emblem**

BELOW: A Bf 109 E-1 of the Second Staffel of an unidentified unit in France early in the Battle. Just visible against the mottling beneath the cockpit is the name 'Mimi', possibly the name of the pilot's girlfriend.

BELOW: Taken in France during June-July, this photograph shows the cowlings 7./JG 2 Bf 109 Es which have been removed prior to the application of the Staffel emblem. Designed by Ofw. Erwin Klee and Lt. Hans Schmidt, the emblem consisted of a thumb pressing down onto a top hat. Note the stencil for the emblem lying on top of the compressed air bottles and spraying equipment in the foreground. The Bf 109 E in the background bearing the single large chevron is believed to be the aircraft in which the Gruppenadjutant of II.Gruppe, Oblt. Adolf Steidle, lost his life near Cherbourg on 11 August.

Messerschmitt Bf 109 E-3 of Oblt. Adolf Steidle Adjutant of III./JG 2
The Bf 109 E-3 of the Adjutant of III./JG 2, Oblt. Adolf Steidle, which was lost on 11 August after crashing near Cherbourg following combat over the Channel with RAF fighters.

"We came face to face and shot at each other simultaneously…"

GERHARD SCHÖPFEL, JG 26

I was appointed *Staffelkapitän* of 9./JG 26 on 23 September 1939. We spent a quiet autumn of 1939 and winter of 1939-40, our *Gruppe's* only *Abschuss* being claimed on 7 November by my friend Joachim Müncheberg from the *Gruppenstab* III./JG 26.

During the campaign against the West, I chalked up four victories. In fact, my first aerial combat impressed itself on me more than my first actual victory. I remember I encountered a Belgian Hurricane in the Beauvechain area. We came face to face and shot at each other simultaneously. What is apparent in such an encounter is that you do not know in which direction your opponent will go and, of course, if you each have the same intention, you could collide. I do not know if I hit him but, in any case, he got me several times in the wings. Fortunately, nothing vital in my aircraft was destroyed.

So, the western campaign ended for me without too much damage and we prepared for the next step. If you ask the question, *did we feel fear before action?* – I would like to tell you the following short story. When I was young, I was a 'pathfinder'. With my gang, we often went to play in ruins. I suffered from vertigo and remained on the ground when my comrades were playing high on the walls. Was it to defeat this fear that I chose to join the *Luftwaffe*? Probably! When I received orders to enlist, I was very anxious and I really wondered if I had chosen the right arm of service. But when I was in action and too busy to worry, all my fear disappeared.

I had the same feeling during my entire career as a fighter pilot. Between the end of the campaign in France and the forthcoming Battle of Britain, I was anxious to know how things would go; but tension of the *Sitzbereitschaft* disappeared as soon the start order was given. Action eliminated all fear.

One of my first victories of the Battle of Britain is a good example of this 'fear-non fear' factor. I had a dog-fight with a British Spitfire at very high altitude. I succeeded in firing at him. The pilot lost control of his aircraft which continued to fly horizontally before diving in spirals and crashing. We were already in France. I overflew the crash site and when I decided to fly away and return to my base, my engine suddenly became troublesome and I had to land immediately. Fortunately, I found a large field and belly-landed without problem. In the proximity, there was a German *Luftwaffe* hospital. A doctor arrived quickly at my landing site and took my heart beat. When he had finished, he looked at me in astonishment and said: "*You just had a dogfight, you've made a belly-landing yet your heart beats as if you were sitting in a chair reading a book – amazing!*"

BELOW AND RIGHT: Hptm. Alexander von Winterfeldt, Staffelkapitän of 8./JG 2 seen in the cockpit of his Bf 109 E 'Yellow 4'. Von Winterfeldt, a pilot in the First World War, was called back to the Reich at the beginning of August in order to take command of III./JG 52, the previous commander, Hptm. Wolf von Houwald having been killed in action. Von Winterfeldt's place was taken over by Oblt. Karl-Heinz Metz.

June-August 1940

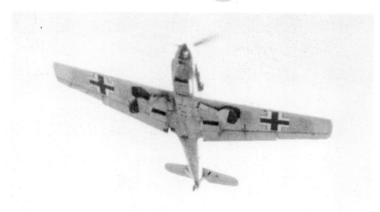

ABOVE: Bf 109 E-1 'Black 12' of I./JG 27 is pictured here shortly after taking off for a mission during the summer of 1940, showing to good advantage the uneven cycling of the undercarriage as it is retracted. Note also that the flaps are still in the partially lowered position to assist in taking off.

RIGHT: A Bf 109 E of 8./JG 2 undergoes maintenance at Evreux-West, August-September 1940. Note the Staffel emblem on the nose – the red 'Springwolf'. This emblem was taken from the family coat of arms of the unit's Kapitän, Hptm. Alexander von Winterfeldt.

8./JG 2
'Wolfhound' emblem

LEFT: A Bf 109 E-4, believed to be 'White 1' of 1./JG 2 undergoes routine maintenance in a camouflaged dispersal bay, probably at Guines during early August.

ABOVE: The Kommodore of JG 2, Harry von Bülow-Bothkamp, prepares for a mission over the Channel in July 1940.

LEFT: A close-up of the JG 2 shield containing the script R of the Richthofen Geschwader. The shield was white with a black border and the 'R' in red.

BELOW: A Bf 109 E-1 possibly of 3./JG 2 seen at Saint Quentin on 27 July 1940.

June-August 1940

LEFT: A Bf 109 E belonging to the Gruppenadjutant of JG 2 seen on the Channel Front in July 1940. Note the aircraft carries the chevron of the Gruppenadjudant which appears to be in the uncommon colour of yellow.

RIGHT AND BELOW: Oberstleutnant Harry von Bülow-Bothkamp, Kommodore of JG 2 seen here wearing the 'Kabok-Schwimmweste' sits in the cockpit of his Bf 109 at the end of July 1940. Note the Script 'R' 'Richthofen' Geschwader emblem.

LEFT: Lt. Hermann Graf climbs out of Bf 109 E White or Yellow '4' in France during the summer of 1940. Graf was posted to 9./JG 52 in July 1940 and went on to become one of the Luftwaffe's greatest aces. Note the unusual position of the First Aid kit symbol which, in this instance, is situated in a compartment behind the pilot's head.

"We fought for Fatherland, not the Führer..."

WALTER STENGEL, JG 51

I was born on 5 July 1907 near Kehl am Rhein and I started flying gliders in 1929. Then, in 1934, I started receiving free training on powered aircraft and I received my pilots' certificate. From 1936, I flew as a reservist in the *Luftwaffe*. In the autumn of 1938, I was commissioned as *Leutnant* in the Reserve. It was only natural that I continued on from glider pilot to the *Luftwaffe*. During the Weimar period, I had been unemployed but I was very interested in sport flying. The National Socialist Party had no involvement in this activity during that time. When war came, we fought for Fatherland not the *Führer*.

I was called to active service on the night of 25-26 August 1939. I went to Karlsruhe where my group met at the railway station and from there we drove to Augsburg. A written order came in from *Luftgaukommando München* transferring us to *E.Staffel* 4 at Fürstenfeldbruck. I had flown the He 70 *Blitz* reconnaissance aircraft but not the Do 17. I had only a B1 certificate. I was supposed to report to an Army reconnaissance unit. I spoke with *Lt*. Rittmeier whom I knew and who got me a transfer to a fighter unit even though I had no fighter training.

In November 1939, II./JG 51 was formed at Eutingen. I was assigned to this unit as a fighter pilot without any fighter pilot training. I was afraid that I would not see any combat before the war was concluded. My *Kommandeur* was *Major* Kramer. He was followed by *Major* Burgaller. We were stationed at Friedrichshafen on Lake Constance during the winter of 1939-40. The edge of the lake was frozen. Burgaller was killed on 2 February 1940 when he crash-landed on the ice and slid into the water below. *Hptm*. Günther Matthes took over the *Gruppe*.

We had strict orders not to cross the French border so as not to provoke the French in any way. There were a few alerts and we took-off to intercept French reconnaissance aircraft but we were too far away from the Rhine and we could not pursue them into France. *Lt*. Josef Priller was the *Kapitän* of the 6. *Staffel*.

My first air battle was during the French campaign over Dunkirk when the British Army was evacuating. On 28 May, we engaged a Hurricane squadron which seemed to consist of beginners. They didn't fly a defensive circle like our other opponents but headed for home. It was the first time that I had fired the guns on my Bf 109. Priller shot down two. I did not score any kills during this campaign. We moved to a different airfield every other day.

After the Battle of France, *Oberstleutnant* Werner Mölders took over the *Geschwader* from *Oberst* Theo Osterkamp. I first met Mölders in Wissant when he took command. I was wearing my flight suit which had been holed during a low-level attack at Chartres several days before when my aircraft had received a hit. Since I did not have a replacement for the ceremony, Mölders had authorized me to wear it.

I flew a total of 505 sorties of which 115 were during the Battle of Britain. I was just the fourth pilot to receive an auxiliary fuel tank carrier. We flew mostly escort missions during the Battle. My most memorable mission was on 29 July, as escort for a Ju 87 formation which attacked a convoy of 22 ships. We had to admire the courage of the Stuka pilots since they flew such a slow aircraft. Our unit sustained the most losses when we escorted Stukas. During the first attack on the convoy, the Stukas scored eight direct hits, destroying their targets. A second attack managed another nine hits. Of the 22 ships, only four came through unscathed. We had our first fatality on this mission when *Fw*. Emmerling who landed his Bf 109 in the *Kanal* off Calais was later found dead.

June-August 1940

Messerschmitt Bf 109 E-3 of Stab/JG 2
The Bf 109 E-3 of Major Harry von Bülow-Bothkamp the Geschwaderkommodore
of JG 2, illustrates an example of the heavily stippled mottle applied to the blue
65 sides of JG 2 aircraft during the summer of 1940.

BEOLW: Oberstleutnant Harry von Bülow-Bothkamp is helped with his life vest
by a member of the ground crew.

ABOVE: Oberstleutnant Harry von Bülow-Bothkamp's Bf 109 E-3 belonging to Stab JG 2 is surrounded by tools, paint pots and engine covers at an unidentified airfield on the Channel Front in late July 1940. The original photograph was marked 'damaged aircraft'.

LEFT: This photograph of ground crew pushing a Bf 109 E-3 of II./JG 2 into one of the hangars at Octeville airfield is believed to have been taken sometime during early August 1940.

Pilots take an alfresco meal whilst a Bf 109 E-3s of 9./JG 2 taxies across Octeville airfield near Le Havre soon after the unit's arrival from Germany.

"Mölders called me and just said: 'Let's shoot them down!'"

ERICH KIRCHEIS, JG 51

Following a short rest after the Western Campaign, JG 51 was posted to the Channel to take on the RAF. I was still *Geschwaderadjudant* which meant I flew as wingman to the *Kommodore*. On 20 July 1940, before the start of the so called Battle of Britain, our 'old' *Kommodore*, *Oberst* Theo Osterkamp received orders to leave the *Geschwader* in the hands of Werner Mölders. At that time, being 48 years of age, Osterkamp was the oldest *Kommodore* to serve in a fighter *Geschwader*. Newly promoted to *Major*, Mölders had been shot down in France with JG 53 and taken prisoner before being released on 30 June. Of course, he was already very well-known and I felt it would be an honour if he kept me as his *Adjudant* and *Rottenflieger*. He took over the *Geschwader* about one week later and kept me at his side.

The first operational mission flying at the head of the *Geschwader* occurred on Sunday, 28 July and nearly ended in catastrophe. We flew towards England. The mission nearly over, Mölders ordered the *Geschwader* to turn for home while he decided to fly some distance from the rest of the unit, perhaps so that he could observe things from a better position. Mölders and I thus found ourselves alone at 7-8,000 metres. We knew that our situation was dangerous; we could be attacked by Spitfires, but even worse was the fact that it could get out that a *Kommodore* had flown over England alone with his wingman. Göring had personally written an order to forbid such a situation; he wanted to preserve his *Kommodore*! When we decided to turn back, we spotted three Spitfires flying 1-2,000 metres below us. Mölders called me and just said: "*Let's shoot them down!*" They had not seen us coming out of the sun and our higher position gave us a good advantage. Mölders approached the enemy aircraft and took up a good position. At that moment, having the task of watching our backs, I spotted more aircraft and warned him: "*Achtung! A Staffel above us and another one in our back!*" Mölders answered: "*Be quiet! They're ours!*" I trusted him: from afar, aircraft are only small dots and are very difficult to identify. The '*Staffeln*' quickly approached us. Now, we saw with horror, that they were not friends. Coming in from a higher position, they opened fire. We maintained direction, approaching the three Spitfires which had not seemed to notice us.

Mölders started shooting and a Spitfire went down. Suddenly, we were surrounded by bullets and were forced to fly away. We dived away in opposite directions. I very quickly dropped down by 4,000 metres, losing my *Kommodore*. I had to make some sharp manouvers and I finally succeeded escaping the British. I landed safely and discovered that my engine had been hit twice.

But I was very anxious to know what had happened to my *Kommodore*. As soon as I was on my legs, I was told that Mölders had already landed, but that he was wounded. His aircraft had been hit several times and he had taken a hit in his left knee. At that very moment, he was on his way to hospital.

Göring had already been informed of his fate. That same day, I was ordered to call him personally. I did it with considerable anxiety and I received the biggest bawling out of my whole life. Fortunately, Mölders was not long in coming back. We all remembered for a long time after that, his first *Abschuss* with JG 51!

RIGHT: Luftwaffe units often went to great lengths to ensure that their aircraft were concealed from the air.

BELOW: View showing a typical camouflaged fighter dispersal area in France in the summer of 1940.

BELOW: Bf 109 Es of III./JG 2 at their wooded dispersal in France in the summer of 1940.

June-August 1940

RIGHT AND BELOW:
Bf 109s of 6./JG 26
receive attention from
the ground crews at
their dispersal at
Marquise – where they
arrived from Germany
on 21 July 1940. Note
the camouflage netting
and the tarpaulins tied
over the lighter
coloured paint on the
aircraft noses which
have probably been
used as a measure of
concealment.

LEFT: This view of personnel of III./JG 26 taking a break in the back of a camouflaged truck is believed to have been taken at Caffiers near the end of July 1940.

RIGHT: After having recovered from wounds, Major Erich Mix retook command of his Gruppe, III./JG 2 on 19 June 1940. He then left France with his unit and relocated to Frankfurt/Rhein-Main on 1 July. On 27 July, he returned to Evreux-West. There, the Gruppe stayed until 4 August. Mix remained Gruppenkommandeur of III./JG 2 until 24 September 1940.

BELOW: The Bf 109 E-4 of Hptm. Erwin Aichele of Stab, I./JG 51 burns out following a forced landing after combat over the Channel on 29 July. Having been born in 1901, Hptm. Aichele, who was killed in the crash, was one of the oldest Jagdwaffe pilots on the Channel Front.

June-August 1940

LEFT AND ABOVE: 1.Staffel, JG 51 suffered its first loss during the Battle of Britain when Ofw. Oskar Sicking was shot down and killed north of Audinghem, France on 20 July 1940. He was killed despite making an attempt to bale out of his stricken aircraft which is seen here lying on the beach following the crash.

ABOVE, ABOVE RIGHT AND RIGHT: A common occurrence, due to the lack of endurance, was that many Bf 109s ran short of fuel and barely made it back to France or had to ditch in the sea. Here another Bf 109 E-3 of JG 51 has made a belly landing on a French beech during a receding tide. German Luftwaffe personnel are trying to salvage the aircraft before the water returns.

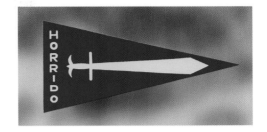

'HORRIDO' emblem of 3./JG 2

Messerschmitt Bf 109 E-4 of 3./JG 2
This unidentified Bf 109 E-4 of 3./JG 2 exhibits two variations from finishes seen on other aircraft of the unit. Finished in an 02/71 scheme with heavy fuselage mottle of the same colours, it carries no JG 2 shield beneath the windscreen and the width of the black borders of the Balkenkreuz have been increased to reduce the visibility of the white segments. The yellow and blue 'Horrido' pennant on the nose, later adopted by Helmut Wick as his personal emblem, was designed by Uffz. Franz Jaenisch, the colours alluding to the Swedish connections of the one-time Staffelkapitän, Hennig Strümpell.

LEFT: Ground crew of 4./JG 52 carry out an undercarriage retraction test on 'White 2 or 12' in France during mid-summer 1940. The Staffel badge of a red cat on a white disc which was carried on the starboard side only, was inherited from 1./JG 71 which, together with 2./JG 71 formed the basis for 4./JG 52 in 1939.

'Red Cat' badge 4./JG 52

THIS PAGE: On 5 August 1940, Hptm. Douglas Pitcairn, Staffelkapitän of 1./JG 51 collided with his wingman, future Ritterkreuzträger, Ofw. Erwin Fleig, on the airfield at Pihen, France during take-off. Despite serious injuries, Pitcairn returned to his unit in October of that year but doctors pronounced that his operational flying days were over. His place was taken by future Eichenlaubträger, Oblt. Hermann-Friedrich Joppien. Note the Staffel's Mickey-Mouse emblem beneath the cockpit on 'White 8'.

LEFT: A Bf 109 E 'Yellow 5' from the 3. Staffel of an unknown Geschwader is towed through a French town some time during the mid-summer of 1940. Note that the retractable undercarriage legs have been tied together to prevent them from collapsing during the journey.

RIGHT: Civilian technicians pose for a photograph with a Bf 109 E-1 of 2./JG 54 as a backdrop probably in France in early August 1940. Note a part of the Staffel emblem is just visible below the cockpit.

LEFT: Bf 109 E 'Yellow 13' of 3./JG 54 in France during August 1940. Note, that unlike other aircraft in the Staffel, this machine does not carry the unit's emblem on the side visible.

June-August 1940

RIGHT: Bf 109 E-3 'Black 2' – probably an aircraft belonging to I.(J)/LG 2 and seen at Pihen-St Inglevert at the beginning of August 1940. It has collided with another aircraft from the unit, probably from 5. Staffel.

BELOW: The 'Flying Clog' emblem was adopted by 7./JG 54 in remembrance of its time in Holland as 3./JG 21 before becoming 7./JG 54 during the Summer of 1940. The clog is white with a black outline and details.

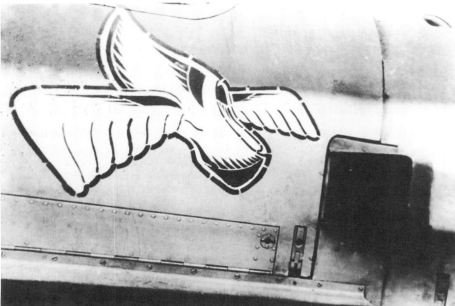

BELOW: At the end of the Western campaign, I./JG 52 was refitted at Nordholz in Northern Germany before moving to Calais on 6 August. Although taken at Calais, the exact date of this photograph is not known, and shows Lt. Hans Berthel of I./JG 52 being welcomed back from a mission by Ofw. Franz Bruden. Lt. Berthel would later become a prisoner of war, being captured after baling out of his crippled Bf 109 on 15 September.

LEFT: Reportedly taken at Guines during August, this view shows 'White 4' and other aircraft of 7./JG 54 in various camouflage finishes dispersed beneath the trees on the perimeter of the airfield.

June-August 1940

RIGHT: The date and location of this photo are uncertain but it shows to advantage the large eagle emblem carried by some aircraft of 6./JG 52. The eagle is blue with white details.

LEFT: Believed to have been taken at Pihen in early August, this photo shows Staffel groundcrew gathered for a photograph beneath the nose of a Bf 109 E of I./JG 51 dispersed beneath camouflage netting on the airfield perimeter.

Werner Mölders

Youth

Werner Mölders was born at Gelsenkirchen/Westfalia on 18 March 1913. At the time of Werner's birth his father, Victor, was working as a teacher in England but with the outbreak of war in August 1914 he was forced to escape home to Germany aboard a neutral Dutch ship. On returning home, he joined the German Army and was subsequently commissioned as a *Leutnant*, only to be killed while serving with *Infantrie Regiment* 145 near Vauquois on the Argonne Front on 2 March 1915 shortly before Werner's second birthday. Following the death of her husband his mother, Anna-Maria, returned to her family in Brandenburg/Havel, faced with the difficult task of raising four children (Hans, Anne-Marie, Werner and Victor) on her own. The Mölders family were devout Catholics but lived in a very strong Protestant environment. Since religion can often offer a form of lonely sanctuary, Werner developed into a very serious boy and would retain a seriousness all his life.

Service in the Army

Werner Mölders as a young Pionierfähnrich talking on a field telephone.

Deciding to follow in his father's footsteps, Werner wanted to become an army officer. Obtaining his 'Abitur' at the age of 17, he enlisted in the small army allowed to Germany by the provisions of the 1919 Versailles Treaty. On 1 April 1931, he joined II./IR 2 at Allenstein in East Prussia. In October 1932, he was transferred to the Kriegsschule at Dresden and to the *Pionierschule* at München in June 1932. With aviation becoming the great dream of many young Germans who remembered the First World War exploits of Bölcke and von Richthofen, the rise to power of the National Socialist Party in 1933 and the creation of a new air force gave Werner the opportunity to transfer to that arm of the services. But, as with his future contemporary Adolf Galland, Mölders would also suffer problems. Whereas Galland's eyes were deficient, Mölders suffered from a fear of heights, a fear that he would conquer with a major effort of willpower.

Service in the Luftwaffe

On 6 February 1934 Mölders joined the DVS (*Deutsche Verkehrsfliegerschule*) at Cottbus and remained there until the end of that year. Following his promotion to *Leutnant* on 1 March 1934, he trained with *Kampffliegerschule Tutow* and *Jagdfliegerschule Schleissheim* until the middle of 1935. On 1 July 1935, he was transferred to *Fliegergruppe Schwerin*, a ground support unit which was later redesignated I./St.G. 162 *Immelmann*. Flying He 45s and He 46s, he was transferred to fighters the following year. Promoted to *Oberleutnant* on 1 April 1936, he led the *Jagdschulstaffel* of II./JG 134 *Horst Wessel* at Werl in Westfalia where his commanding officer was *Major* Theodor Osterkamp, a veteran of the First World War, credited with 32 aerial victories. On 15 March 1937 Mölders took command of 1.*Staffel* of I./JG 334 at Wiesbaden and his unit, equipped with the Heinkel He 51, would be successively redesignated I./JG 133, and then I./JG 53 *Pik As*.

The Spanish Civil War

The seriously-minded Mölders was still a bachelor when he was sent to Spain in May 1938. On the 24th of that month, he succeeded Adolf Galland as *Kapitän* of 3./J 88. This was the first time that the paths of the two men crossed. At the same time, the obsolete He 51s were replaced by the new Bf 109 *Dora* which would later be replaced by the *Emil* becoming the best fighter used by either side during the Spanish Civil War. Combining his own abilities with the qualities of the Messerschmitt fighter, Mölders quickly achieved success and in his first aerial engagement, shot down an I-15. Four days later, two further victories were added, another I-15 and an I-16. With the exception of an SB-2 shot down on 23 August 1938, Mölders would claim only Polikarpov fighters until the end of his stay in Spain. On his return to Germany on 5 December 1938, he was credited with 14 victories plus an additional three that were unconfirmed. Promoted to *Hauptmann* as the highest scoring German ace of the Spanish Civil War, he was then temporarily assigned (as had been Galland earlier) to the Air Ministry to study and improve fighter tactics based upon experiences gained during the Spanish conflict. His influence was to be enormous in that he proposed the deployment of a loose formation of four aircraft – the '*Schwarm*' – broken up into two elements of two – the '*Rotte*'.

The 'Sitzkrieg'

While Galland was transferred to II.(*Schlacht*)/LG 2 equipped with the Hs 123, Mölders returned to his old fighter unit to lead 1./JG 53 (formerly-1./JG 133). It was during this time that he acquired his nickname of '*Vati*' ('Papa') due to his serious nature, experience and rigidity. This nickname was not intended as offensive but one which was born out of respect. No-one feared Mölders and he was very popular amongst his pilots. He was not an impetuous man and could drink a glass of beer like the rest - but never two!

If his successes in Spain were partly due to his good fortune in receiving the best aircraft of its time, then the *Sitzkrieg* – or Phoney War – was to prove that he was an excellent fighter pilot and tactician. His introduction to the new campaign was, nevertheless, quite unsettling. On 8 September 1939, he led three other Bf 109s in an attack on six French Curtiss H-75s of GC II/4 north of Karlsruhe. In the ensuing dogfight, Mölders' Bf 109 was heavily damaged, forcing him to crash land in a field near Wölfersweiler. Trapped in his cockpit and slightly wounded, he had to wait for a local *Flak* crew to release him. Strangely, the French pilots involved claimed two victories, attributed to three pilots (S/C Cruchant being credited with two claims combined with two other pilots)!

Mölders recovered quickly and claimed his first victory over the border twelve days later. Taking off with his *Schwarm* to Trier, he destroyed another H-75 of GC II/5 from a patrol escorting a reconnaissance aircraft. Sgt Quequiner, piloting N°21, was able to bale out of this aircraft which crashed near Merzig.

After being promoted *Kommandeur* of III./JG 53, Mölders celebrated his new command by shooting down a Blenheim I (L6694) of No. 57 Sqn engaged in reconnaissance along the Moselle on 30 October 1939 but would have to wait until 22 December to obtain his third victory in France. While escorting a Do 17 P of 1.(F)/123, he attacked some fighters identified as "Moranes" but which were in fact, Hurricane Is of No. 73 Sqn RAF. With his wingman, *Oblt.* von Hahn, he shot down two (L1967 and N2385) near Budange. With the onset of bad weather, the first months of 1940 were quiet but on 2 March, at the end of a very scrappy encounter, *Hptm.* Mölders and *Uffz.* Neuhoff were able to claim two Hurricanes (L1808 and L1958) from No. 73 Sqn which crashed near Metz. The following day, again around Metz, Mölders engaged a Morane Saulnier 406 of GC II/3. This was claimed destroyed but, in fact, C/C Koerber, although wounded, managed to land his damaged aircraft at Toul airfield. On 26 March, another MS 406 was claimed near Trier, but this proved to be a Hurricane of No. 73 Sqn whose pilot, F/O Edgar James 'Cobber' Kain of the RNZAF, baled out after having previously been shot down on 2 March! On 2 April, another Hurricane, this time from No. 1 Sqn, was shot down near St Avold but the pilot was able to force-land his heavily damaged fighter behind the Allied lines and avoid capture.

On 20 April, III./JG 53 were flying in the Zweibrücken area where they encountered Curtiss H-75s of GC II/4 escorting a Potez 63.11 reconnaissance aircraft of GR II/36. In the combat that ensued, anti-aircraft guns shot at both sides! An H-75 N°136 fell to Mölders while another was damaged by *Flak*. The pilot, C/C Cruchand, was seriously wounded but managed to crash-land his fighter near Biesbrück. On 23 April, Mölders claimed his last victory of the *Sitzkrieg* when he shot down a Hurricane I (N2391) of No. 73 Sqn during the morning near Sierck-les-Bains, the pilot, Sgt C. Campbell parachuting to safety. During this campaign, *Hptm.* Mölders was credited with nine additional victories while Adolf Galland flew only ground support. By the time Galland did transfer to the fighter arm, Werner Mölders had 23 official victories.

The Campaign in the West

On 10 May 1940, III./JG 53 was based at Wiesbaden airfield and Mölders had to wait four days before he was credited with his first victory during the invasion of the West, this being a Hurricane on the 14th of the month. During the first days of the attack and mainly over France, III./JG 53 had to escort the bombers and were ordered not to attack enemy fighters. On 15 May,

Werner Mölders at the celebratory dinner after receiving the Ritterkreuz on 29 May 1940 having achieved 20 aerial victories at this time.

another Hurricane was claimed by the *Kommandeur,* but it would be a French cockade that was later painted on the rudder of all his aircraft to record that particular victory. On 17 May, III./JG 53 was transferred to Douzy, near Sedan in France from where the unit flew air cover sorties over the *Wehrmacht* spearheads advancing near Cambrai. On 19 May, Mölders was credited with a 'P-36' (almost certainly a Bloch 152, which was often confused with the Curtiss). During the evening of the 20th, Mölders claimed his 13th victim, a British bomber described as a 'Wellesley'. On 21 May, three MS 406s were shot down (apparently aircraft from GC I/6 and III./6) and on the 22nd, it was the turn of a Potez 63.11, N°315 of GAO I/514, shot down near Montagne de Reims. Another Morane was lost during the evening of 25 May (Mölders' 18th victory) and on the 27th, two Blochs, thought to have been from GC I/8, were claimed south of Amiens.

The two young Geschwader-kommodore, Werner Mölders and Adolf Galland in discussion with Reichsmarschall Hermann Göring at an unidentified location in France during the summer of 1940.

With 20 victories over France and 14 in Spain, Mölders was awarded the *Ritterkreuz* which was presented to him on Loe airfield, near Le Selve. On 31 May, near Abbeville, Mölders shot down a LeO 451 of GB 1/12. On 3 June, during Operation *Paula* (launched primarily as a propaganda operation), Mölders claimed two victories – a Curtiss H-75 (which, in fact, was a Bloch 152, and which was subsequently identified on his rudder with a British roundel!) and, very unusually, a Spitfire. Exactly, what a Spitfire was doing near Paris at a time when all RAF units had retreated to their bases in England to fight over Dunkirk is unclear. The 'Spitfire' was probably a D.520 of GC I/3. Two days later, Mölders experienced altogether different circumstances. At around noon, he was credited with the destruction of a Bloch 152 (N°651 of GC I/8?) and a Potez 63.11 (N°250 of GAO 501?) and later that afternoon, whilst on his second mission of the day, he spotted some "Moranes" attacking some Bf 109s. He decided to intervene but the "MS 406s" turned out to be potent D.520s of GC II/7. Having under-estimated the enemy type, Mölders was shot down by S/Lt René Pommier Layrargues, his Bf 109 E-3 crashing near Canly. Mölders was able to parachute to safety, but was captured on the ground by soldiers of 195e RALT, an artillery unit who set upon him before an officer intervened. Interested in the man who shot him down, Mölders asked to meet him, only to find that Pommier Layrargues was already dead, having been brought down and killed at Marissel a few minutes after their engagement.

Mölders ended the *Westfeldzug* in a French POW camp at Montferrand. With the fall of France, he was eventually freed at the end of June 1940 and this is where there is cause for some interesting speculation! If he had been captured by the British in May, he would almost certainly have been sent to a POW camp in Canada, ending the war in safety and terminating the career of a great pilot. But as a prisoner of the French, he was liberated and became – posthumously – a flying legend. Which was the better fate?

This Bf 109 E-4 was presented to Mölders by the people of the Saar mountains. The inscription on the nose reads 'Saarbergmann Glück auf' ('With luck from the people of Saar').

The Battle of Britain

After a short period of leave, Mölders, promoted to *Major* on 19 July, returned to III./JG 53. Soon afterwards, however, he left to take over command of JG 51. At that time, Adolf Galland was appointed to lead III./JG 26 after having shot down 14 planes in the *Westfeldzug* whilst with JG 27.

As is often the case, establishing a new command proved hectic for Mölders. On 28 July, the new *Kommodore* damaged a Spitfire I (P9429) of No. 41 Sqn, RAF. Wounded in the thigh, the pilot, F/O A.D.J. Lovell, managed to land his damaged aircraft at Hornchurch. F/O Lovell survived to become an ace in his own right, only to be killed in a flying accident in 1945. Shortly afterwards, Mölders himself was shot down by F/Lt John Webster of the same Sqn. This was Webster's fifth claim but he was killed on 5 September 1940 when his parachute failed to open after baling out following a collision with another Spitfire of 41 Sqn. (*Author's note: another source attributes this claim to the ace, 'Sailor' Malan of 74 Sqn - see page 17*). Wounded in the knee, Mölders was able to force-land his damaged Bf 109 on the French coast. He returned to his unit on 7 August, but would have to wait some time before he could fly again.

On 26 August 1940, Mölders submitted his 27th claim, another Spitfire. By 20 September, his score had reached 40 enemy aircraft shot down, proof that the battles over England were very intense and on that day, he was was credited with two more Spitfires (X4417 and N3248) of No. 92 Sqn and was awarded the Oak Leaves to his *Ritterkreuz*. He was only the second member of the German armed forces to receive the decoration. Four days later, Adolf Galland also received the award, becoming the third person to do so. It was at about this time that German newspapers devised a kind of competition between the two aces. One publication would be 'for Mölders' another "for Galland"; in reality however, Mölders was not interested in such "competition". He told Galland: "*In this war, you will be the Richthofen and I the Bölcke*" - yet further proof that the serious *Kommodore* was more interested in tactics than glory.

Mölders is seen here exiting the cockpit of his Bf 109 E-3 during the latter part of August 1940.

Mölders score continued to increase; on 27 September, it was a Spitfire over Kent, possibly P9364 of No. 222 Sqn. piloted by Sgt Ernest Scott, who was killed after having shot down a Bf 109 – his fifth confirmed victory. On 11 October, another Spitfire I went down (X4562 of No. 66 Sqn) and next day, three Hurricane Is (P3896, V7251 and V7426) of No. 145 Sqn. On 17 October, Mölders claimed another Spitfire (R6800 LZ-N of No.66 Sqn.) followed by three more Hurricanes on 22 October (possibly from Nos. 46 and 257 Sqns) off the English coast. Mölders now had his fiftieth victory. Galland reached this total eight days later. From the beginning of October, Mölders became the

Sitting in the cockpit of his Bf 109 F, Werner Mölders is seen here describing another sortie.

first pilot to test the new Bf 109 F in combat, which soon proved superior to contemporary British fighters. Certainly, this also helped in his subsequent successes.

After spending a few days leave skiing, JG 51's *Kommodore* returned to action at the beginning of 1941. Exploiting the relative inactivity of the *Luftwaffe* in the west (the German High Command was preparing to attack the Soviet Union and had moved many units to the east), the RAF were beginning to conduct sorties over France and the fighting now took place mainly off the French coast. On 20 February, Mölders claimed two Spitfires (his 57th and 58th victories). Five days later, a Spitfire II (X4592 of No. 611 Sqn) was shot down, and on the following day he scored his 60th victory. Galland had to wait until 15 April to attain the same score.

On 13 March, Mölders shot down another British ace, S/Ldr Aeneas 'Donald' MacDonnel. MacDonnel, from No. 64 Sqn, was born in Baku in 1913, and was the 22nd Hereditary Chief of the Glengarry Clan. Leading a sweep over Northern France, MacDonnel (credited with nine or ten victories) was shot down by Mölders (his 62nd victory) and baled out into the Channel. He was rescued by a German motor boat but remained a prisoner of war until 1945.

Mölders in the cockpit of his Bf 109 F. One of the first members of the Jagdwaffe to take the new 'Friedrich' into combat in October 1940, this photo may well be from that period.

The new versions of the Hurricane and Spitfire proved no match for the Bf 109 F. This is well indicated by a list of Mölders's claims for the period:

15 April   Hurricane II of No.615 Sqn
16 April   two Hurricane IIs of No.601 Sqn
                 (one claimed as a "Spitfire")
4 May   Hurricane II (Z3087) of No.601 Sqn
6 May   Hurricane II (Z2743) of No.601 Sqn
8 May   Spitfire II of No.92 Sqn.

Mölders' aerial victories declined following the transfer of JG 51 to the East. On 21 June, Adolf Galland – then with 69 claims – was the first *Luftwaffe* pilot to add the Swords to his *Ritterkreuz*. On the eve of *Barbarossa* – the German invasion of the Soviet Union, Mölders had 'only' 68 claims, but on the day of the invasion, he claimed an I-153 (which must have brought back memories of Spain!) and three SB-2s shot down. He was awarded the Swords but this time as the second pilot to receive the decoration.

The rudder of Mölders' Bf 109 F (note rounded starboard wing tip in background) displaying 54 'Abschuss' bars indicate that this photo was taken during early to mid-October 1940.

At this time, Soviet aircraft and pilots were seen as generally inferior to their German counterparts and this enabled Mölders and his men to claim unprecedented scores and on 30 June, he was credited with the destruction of no fewer than five enemy aircraft. By 15 July 1941, on his 291st combat mission he claimed his 100th and 101st victories and was awarded the Diamonds to his *Ritterkreuz*. By comparison, Galland, would have to wait until 28 January 1942 for this decoration.

By this time, Mölders had achieved an almost mythical status, seen to be deserving of 'protection'. He was ordered not to fly ('*Flugverbot*') to avoid risking his life at the front and was transferred to the Air Ministry in Berlin. On 7 August 1941, he was promoted to Inspector of Fighters and left his unit and on 13 September 1941, he married Louise Baldauf, the widow of a fallen comrade.

Mölders could have remained safely at the Ministry, close to his wife, but he was preoccupied with the Soviet campaign and visited the Eastern Front many times. In the autumn of 1941, he went to the Crimea to lead the combined operations of *Stukas* and fighters where he discovered an important supply problem which he tried to resolve. In spite of the *Flugverbot*, he wanted to have a clearer picture of the situation in the air by flying again. On 8 and 11 November, Mölders borrowed a Bf 109 of III./JG 77 and shot down three more Soviet aircraft over Sevastopol and the Kertsch peninsula, though he did not record them officially. Future *Ritterkreuzträger*, Herbert Höhne, remembered serving as Mölders' wingman at this time. After spotting enemy aircraft, the Inspector led his *Kaczmarek*, giving him instructions by radio and 'donating' him his victories. It would seem that '*Vati*' Mölders enjoyed the role of 'counsellor' and adviser.

Oberstleutnant Mölders as photographed by his wife at his desk at the Air Ministry just before leaving for the Eastern Front.

On 17 November 1941, *Generaloberst* Ernst Udet committed suicide and Mölders was called back to Berlin to

Enjoying himself skiing during a period of leave during the winter of 1940 – 41, Werner Mölders (centre) is seen here with a group of his collegues at an Alpine resort.

assist with the funeral. Four days later, he began his journey to the capital as a passenger in a He 111 of III./KG 27 piloted by *Oblt*. Kolbe, another former flyer from Spain. The weather was bad and following an interim stop at Lemberg, the Heinkel took off again but the weather conditions continued to deteriorate. Near Breslau, the port engine failed and the crew tried to land at the nearest available airfield, Schmiedefelde. At low altitude, the second engine cut and the He 111 '1G+TH' hit the ground near Martin Quander Farm at N°132 Flughafenstrasse. Mölders was killed at 11.30 on 22 November. He was succeeded as Inspector of Fighters by Adolf Galland.

As is often the case after a plane crash (Balbo, Sikorsky, Todt, etc.), rumours circulated in some quarters about a plot to kill Mölders but post-war research has found these to be totally without foundation. It is true that Mölders, as a devout Catholic, criticised the Nazi Party many times for its activities against the church. But to kill Germany's greatest ace for such beliefs at such a critical period in the war is, in the author's opinion, inconceivable.

Werner Mölders was buried in the *Invalidenfriedhof* at Berlin where Manfred von Richthofen already lay. His *Geschwader*, JG 51, later adopted the honour name '*Jagdgeschwader* Mölders'.

As a postscript to this biography it is worth quoting the words of another ace, Dietrich Hrabak: *"Wir waren nur Jagdflieger. Mölders was mehr als das!"*: "We were only fighter pilots. Mölders was more than that!".

ABOVE LEFT: Mölders (left) seen here with the Commander of Panzergruppe 2, Generaloberst Heinz Guderian (right).

ABOVE: Oberst Mölders seen during his tenure as Inspekteur der Jagdflieger during a tour of the Crimea in the Autumn of 1941.

Mölders' funeral procession where, in true military fashion his coffin is carried on a gun carriage flanked by a guard of honour. Following immediately behind is Herman Göring who in turn is followed by five of the most highly decorated Jagdflieger.

June-August 1940

LEFT: Bf 109 E-1 'White 7' of 1./JG 27 in flight over France in August 1940. Note the yellow engine cowling, the Gruppe emblem and the two-coloured spinner, with the tip in the Staffel colour of white.

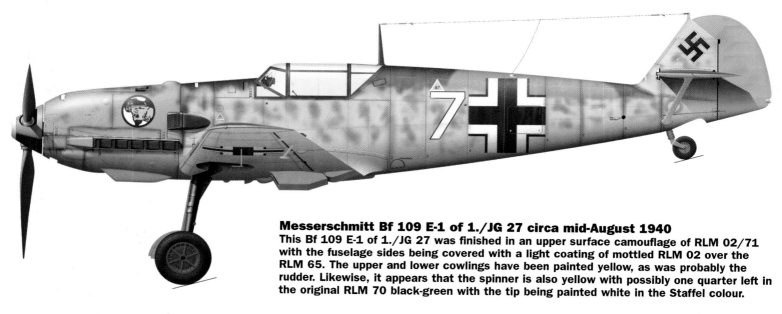

Messerschmitt Bf 109 E-1 of 1./JG 27 circa mid-August 1940
This Bf 109 E-1 of 1./JG 27 was finished in an upper surface camouflage of RLM 02/71 with the fuselage sides being covered with a light coating of mottled RLM 02 over the RLM 65. The upper and lower cowlings have been painted yellow, as was probably the rudder. Likewise, it appears that the spinner is also yellow with possibly one quarter left in the original RLM 70 black-green with the tip being painted white in the Staffel colour.

RIGHT: An unidentified Bf 109 E-4 of III./JG 51 believed to have been photographed in France during the early summer of 1940. Just visible beneath the cockpit is the emblem of III./JG 51, 'Axt von Niederrhein'.

ABOVE: Ofw. Erwin Leykauf climbs into the cockpit of 'White 12' at Guines sometime in early August 1940. As can be seen in this photograph, Leykauf is wearing ordinary shoes rather than flying boots. This was not a uncommon practice as, in the event of the piloy ditching, it was easier to remove shoes than the more cumbersome flying boots.

RIGHT: In this photograph, Erwin Leykauf has now settled in the cockpit of 'White 12'. Of note are the small proportions of both the aircraft number and fuselage Balkenkreuz.

"It was a wonder that I was not shot at..."

SIEGFRIED BETHKE, JG 2

I was born on 24 June 1916 in Strassen, Pomerania. My mother's brother, Erich Voss, was in the *Luftwaffe* during the First World War but he was shot down and killed in 1918 whilst serving as an aircraft observer. In 1934, there was high unemployment in Germany. At the same time the military began to expand. There was also no opportunity for me to find higher education so as to develop a career. I was interested in becoming a pilot due to the fact that my uncle had been in the air force. At the beginning of 1935, after I had taken my *Abitur*, I went into the Navy to become a naval pilot. I reported to the naval school in Flensburg. Then I was transferred to the *Luftwaffe* in 1936 where I completed the A, B, and C (blind-flying) courses at Celle.

My first unit assignment was at Bad Aibling. The unit later was redesignated JG 51. Douglas Pitcairn was my *Staffelkapitän*. I was in his *Staffel* for one and a half years. Shortly before the war started, I was transferred to Herzogenaurach. Then I went to Fürstenfeldbruck to an *Ergänzungsstaffel* with which I went to Merseburg. On 1 May 1940, I was assigned to 2./JG 2, the *Staffelkapitän* of which was *Hptm*. Karl-Heinz Greiser who had fought in Spain.

During the French campaign, I scored four kills. My first kill was on 14 May near Sedan. I attacked a Morane from behind and it exploded under my two cannon. The other three kills were a Potez 63 (in our archive it is listed as a Blenheim on 25 May at 20.35), a Spitfire near Calais (on 26 May at 09.40), and a LeO 451 (in reality a Douglas DB 7 at 20.10). After shooting down the latter, I attacked a second LeO 451 and was myself shot down over Amiens. The combat had taken place on 31 May at low-level and I was shot down either by the gunner of the LeO or by *Flak*. I bailed out. The LeOs had attempted to attack German troops. I landed between the lines and I was rescued by German infantrymen. On the other side were coloured French colonial troops and it was a wonder that I was not shot at. I had to go to a hospital in Köln since I sustained concussion and I was supposed to stay there ten days but I returned to my *Staffel* early on 4 June without the doctor's permission! My unit's doctor hadn't given me permission to fly since my injury had not healed. The unit moved from one base to another until the conclusion of the campaign.

We deployed to the area of Rouen. I was promoted to *Staffelkapitän* when the previous commander took over a *Gruppe*. I commanded the *Staffel* for two and a half years until October 1942.

During the Battle of Britain, I shot down Spitfires and Hurricanes. On my first mission over England, on 11 August, I shot down two Hurricanes over Southampton at 4,000 metres. I attacked them from behind and the first Hurricane immediately exploded and I had to turn to avoid the debris. The pilot of the second Hurricane bailed out. My *Staffel* reported two other Hurricanes shot down during that clash, one being claimed by the future *Schwerterträger*, *Fw*. Erich Rudorffer.

June-August 1940

RIGHT AND BELOW: A Bf 109 E-1 of an unidentified Jagdgeschwader, possibly 3./JG 2, probably seen in France during the summer of 1940. The aircraft is probably 'Yellow 8' and – as is evident from the damage to the cockpit and the mud on the landing gear – it probably made either a 'Fliegerdenkmal' or a belly-landing.

Thumb and Hat emblem of 7./JG 2

Messerschmitt Bf 109 E-1 of 7./JG 2
'White 5' of 7./JG 2 is finished in an upper surface scheme of RLM 02/71 with the sides of the fuselage and fin and rudder covered in a fairly dense mottle of the same. It carries the earlier style III.Gruppe symbol aft of the Balkenkreuz and on the cowling, the 7.Staffel 'Thumb on a Top Hat' emblem designed by Leutnant Hans Schmidt and Oberfeldwebel Erwin Klee.

LEFT: Bf 109 E-1 'White 5' of 7./JG 2 gets airborne from one of the airfields in the Pas de Calais during the early summer of 1940. Visible on the cowling is the 7.Staffel 'Thumb on a Top Hat' emblem.

June-August 1940

LEFT: A Schwarm of 5./JG 27 Bf 109 E-1s during the summer of 1940. While three of the aircraft have mottling applied to their cowlings, aircraft '10' in the foreground does not, suggesting either a high demarcation Blue 65 or possibly yellow cowling.

BELOW: Groundcrew replacing a Daimler-Benz 601 engine change for a Bf 109 E-1 of 8./JG 2 at Octeville in early August.

June-August 1940

LEFT AND BELOW: Two photographs showing a visiting He 111 landing at JG 2's airfield at Octeville in early August 1940. Of note on the JG 2 Bf 109 in the foreground, is the heavily applied 'stipple' mottling over the RLM 65 on the fuselage side and the motor car type rear view mirror mounted on the canopy framework.

LEFT: Along with members of his groundcrew, Gefr. Josef 'Jupp' Bigge is seen here sitting on the port wing of his Bf 109 E-1 'Black 9' of 8./JG 2 at Octeville near Le Havre in early August 1940. The clearly visible 'Springwolf' Staffel emblem was taken from the family coat-of-arms of the Staffelkapitän, Hptm. Alexander von Winterfeldt who by this time, had been promoted to take command of III./JG 52, his position as Staffelkapitän taken by Oblt. Karl-Heinz Metz.

June-August 1940

**9./JG 2
'Stechmücke'
(Mosquito) emblem**

LEFT: Bf 109 E-1 'Yellow 8' of 9./JG 2 is believed to have been the aircraft of Lt. Rudolph Rothenfelder, the designer of the Staffel emblem which consisted of a black and white 'Stechmücke' (Mosquito) superimposed on a black bordered yellow disc.

Messerschmitt Bf 109 E-1 of 9./JG 2
'Yellow 8' of 9./JG 2 is believed to have been the usual mount of Lt. Rudolph Rothenfelder, the designer of the Staffel emblem which consisted of a black and white 'Stechmücke' (mosquito) superimposed on a black-bordered white disc. The aircraft is thought to have been finished in a high demarcation of RLM 70/71 scheme with heavy fuselage mottling of 02 and 71 and carries the earlier III. Gruppe symbol aft of the Balkenkreuz in yellow with a thin black border. In addition to the Staffel emblem the JG 2 shield was carried beneath the windscreen on both sides of the fuselage.

LEFT: Oblt. Helmut Wick photographed in early August while describing a dogfight to his Kommandeur, Major Hennig Strümpell. Ending the French campaign with 14 claims, Wick was promoted to command 3./JG 2 on 23 June 1940.

June-August 1940

RIGHT: A pilot of 8./JG 2 at readiness with members of the ground crew, at Octeville, at the beginning of August 1940. Note the wavy line fuselage marking denoting a III. Gruppe aircraft. The aircraft in the background carrying white markings is from 7./JG 2.

LEFT: Pilots of 8./JG 2 relax while at readiness on the field at Octeville near Le Havre at the beginning of August 1940.

BELOW: Gefr. Josef 'Jupp' Bigge of 8./JG 2 sits on the wing of his Bf 109 E at Octeville in 1940. Following its location at Evreux, the Gruppe transferred to Octeville. Bigge was posted to 8./JG 2 on 20 June 1940. Note the old type of 'Kabok-Schwimmweste'.

LEFT: A Rotte of Bf 109 E-3s of II./JG 2 off of the coast of France fly southwards down the Channel during the summer of 1940.

BELOW: Bf 109 E 'White 7' of 1./JG 51 with its wings dismantled on the French coast, August 1940. Compare with the aircraft seen on page 46.

I./JG 51 'Kitzbüheler' emblem

1./JG 51 'Mickey Mouse' emblem

Messerschmitt Bf 109 E-3./JG 51

This Bf 109 E-3 of 1./JG 51, 'White 7' is finished in an upper scheme of 02/71 with a fairly heavy fuselage mottle of both upper colours. Beneath the windscreen can be seen the I./JG 51 'Kitzbüheler' goat emblem while beneath the rear of the canopy is the 1. Staffel emblem of a stylised running mouse carrying a revolver. A Luftwaffe POW belonging to this Gruppe who was interrogated at this time, stated that this unit was known as the 'Gemsbock

June-August 1940

II./JG 54
'Lion of Aspern' emblem

LEFT: The 'Lion of Aspern' emblem of II./JG 54 originated with I./JG 134. It was adopted by II./JG 54 via I./JG 76 during 1939-1940. The emblem featured a black lion with red and white details on a white background with black border. The lower section of the background shield is red with a white cross.

LEFT: Pilots of III./JG 2 play with a canine friend whilst on readiness at a French airfield in early August 1940.

RIGHT: 'Black 1', a Bf 109 E-4 of 8./JG 26 is seen here jacked up for routine servicing on a summer evening, probably at Caffiers during mid August 1940.

ABOVE: The emblem of 8./JG 26 (white with black details and outline) – based on the 'Adamson' cartoon character and seen below the cockpit of one of that unit's Bf 109s.

**8./JG 26
'Adamson' emblem**

ABOVE: Gefr. Heinz Zimmer is seen here posing beside one of the machines 'Black 9' of 8./JG 26 which displays the Staffel 'Adamson' cartoon emblem.

BELOW: An adapted version of the 'Adamson' Staffel emblem adorns the sign leading to a latrine at 8./JG 26's base at Caffiers. The intended use of 'The Times' can only be speculated!

June-August 1940

"My aircraft rolled over and went into a spin…"

RUDOLF ROTHENFELDER, JG 2

I was born on 7 November 1918 at Kaufering. In September 1939, I was a member of 1./JG 20 under the command of *Oblt.* Walter Oesau with *Major* Siegfried Lehmann as *Gruppenkommandeur* of I. *Gruppe*, soon succeeded by *Hptm.* Hannes Trautloft. On 15 March 1940, several pilots were posted to the newly-created III./JG 2 under *Major Dr.* Mix. Together with my friend Peter Neumann-Merkel, I was posted to 9. *Staffel* under the command of *Oblt.* Hannes Rödel. I designed the *Staffel* emblem – a stinging fly on a yellow background which was our *Staffel* colour.

Following the successful campaign in the West, we left Evreux-West in France on 1 July 1940 to go to Frankfurt Rhein/Main where we stayed until 26 July. We then returned to Evreux from where we started our 'Battle of Britain'. On 29 July, our Geschwader moved to Le Havre-Octeville from where we would fly against England. The *Geschwaderstab* together with the I. and II. *Gruppen* moved to Beaumont-Le-Roger where there were intensive preparations for the offensive. On 4 August, our *Geschwader* was ready for action.

Equipment for our pilots was well-suited for the environment. We received a number of items which enabled us to survive over the Channel Front. First, there was the yellow *Kabok* life jacket which was later replaced by an inflatable life vest. A flare pistol with the appropriate flares was attached to our leg and was to assist us by making us more observable if we went into the 'drink'. Supplementing this was the yellow-dye pouch, which would make it easier to spot us in the water from the air. Special attention was given to the pouch which contained our emergency sea-rations. Besides chocolate with caffeine, Pervitin tablets, and various other items, there was a small bottle of French cognac which needed to be continually replaced! At first, flying over the Channel from Le Havre to the Isle of Wight was considered very difficult but we took measures to counter this fear. Before our fighters took-off, our '*Nebelkrähen*' ('smoke crows'or pathfinders) lay down a smoke trail using smoke cannisters which our fighters could follow. This enabled us to get our bearings, especially so when one of our comrades went into the water. We could then report to the Air Search and Rescue as to what number cannister was the closest. The greatest disadvantage of course, was the fact that the British would also see the smoke and they would know for certain when we would arrive and the course which our formations would be flying. It wasn't long before we had to relinquish the services of our '*Nebelkrähen*'. For sea rescue we used other measures. Neither the so-called 'Udet-buoys', nor the floats did the trick. Despite the prominent display of the Red Cross on their wings and fuselage, they were shot-up by the British. The best means of survival turned out to be the rubber dinghy which was dropped by parachute and the good old life jacket, not to mention the other rescue equipment. Last but not least, we trusted our comrades who reported our location on their return flight. They did everything humanly possible to assure that every means was taken to successfully rescue us.

I must sing a song of praise to our comrades in the Navy and their rescue boats. These were small, fast boats which were always ready to go into action when we were flying against England and many of our comrades thanked them for saving their lives. The same praise must also go to the crews who were active in the Air Search and Rescue Service who flew the flying boats, the He 59 and the Dornier *Wal*.

The period 4 –10 August was used to acquaint ourselves with the different conditions of the Channel Front and the North Sea and to keep the Channel under surveillance. In between these sorties, we were always at readiness to take-off which was a policy which had been in effect since the start of the war and which was co-ordinated between the individual *Staffeln* of the *Gruppe*. Occasionally, we had 'visitors from the other side' which triggered an immediate emergency take-off, but these were often without success. With a feeling of great suspense, we awaited our first operational orders to fly over to the 'opposite side'. To our surprise, on 10 August 1940, as we gathered together and chatted in our excellent accommodation, our squadron leader, Hannes Röders, was called to a briefing by the *Kommandeur* that evening at 21.30. We awaited with great anticipation for his return and it turned out to be true—the time had come. Right up until midnight, we discussed the details of the forthcoming mission. We would first fly at low-level to transfer to Cherbourg-Théville where other single- and twin-engine fighter units would assemble before flying in the direction of the Isle of Wight at noon. The operation was 'Top Secret'—the news was even to be kept from our ground personnel.

Everything happened the next morning—we took-off at 08.06 hours from Le Havre and we arrived at Cherbourg-Théville at 08.45 hours. Units came from every conceivable direction—the I. and II. *Gruppen* including the JG 2 *Geschwader* Staff, our comrades from the *Mölders Jagdgeschwader* and several twin-engined *Zerstörer* units landed at the base. After receiving final instructions, we took-off on our first mission against England at 1055 hours on 11 August 1940. It was a truly uplifting sight to see what we had assembled in the air. They came not only from Cherbourg but also the other airfields in the region. There were 400-500 fighters and *Zerstörer* (Me 109 Es and Me 110s) which were flying at various altitudes towards the Isle of Wight. Our *Staffel* flew top cover at about 9000 metres and we could see, in the distance, that our fear that the British would not come up to engage us was groundless. Over Portland there was already an intensive air combat in progress and we could see the first parachutes drifting to earth. In the water below, we recognised green patches which indicated our pilots. The enemy had sent up 500 aircraft—Spitfires, Hurricanes and Beaufighters—which were now engaged in wild dogfights with us.

I was flying in a *Rotte* element with *Oberleutnant* Fricke, who was flying his first combat sortie, when we were attacked from above and out of the sun by six Spitfires. Over the radio, I warned my comrade who unfortunately failed to react. As the Spitfire was getting into position behind me, I pulled up at full throttle and my aircraft rolled over and went into a spin right into the mass of fighting aircraft until I reached about 6000 metres, where I was able to bring my aircraft under control and pull out of the dive. I saw Fricke's aircraft burning and diving towards the earth but he was unable to bail out.

The first combat had to be broken off when we were low on fuel. *Oblt.* Röders and *Lt.* Kluge had each shot down a Hurricane. As I was flying back to base, I caught a Spitfire which went into the sea burning but unfortunately I did not have a witness.

The first battle of the campaign over Portland was very successful for us. According to our own reports, we had scored 153 kills and we had lost 48 aircraft. I cannot vouch for the accuracy of these statistics but they must be fairly accurate.

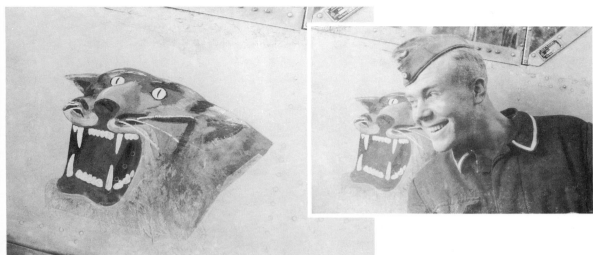

RIGHT: A member of the ground crew smiles broadly next to the tiger's head emblem of 4./JG 26. See Classic Colours, Volume One, Section 4, pg. 300 for variations of this emblem.

ABOVE: The emblem of 4./JG 26 – a yellow tiger's head with red, white and black details.

III./JG 54 Emblem

Messerschmitt Bf 109 E-4 of Stab III/JG 54

This Bf 109 E-4, flown by Oblt. Albrecht Drehs, force-landed at Hengrove near Margate, Kent on 12 August. Upper surface camouflage is RLM 02/71 with a randomly applied mottle to the sides of the fuselage and fin and rudder in a style common to JG 54. The III./JG 54 emblem was carried on both sides of the cowling and the Stab Winkel and bar were white with a thin black border.

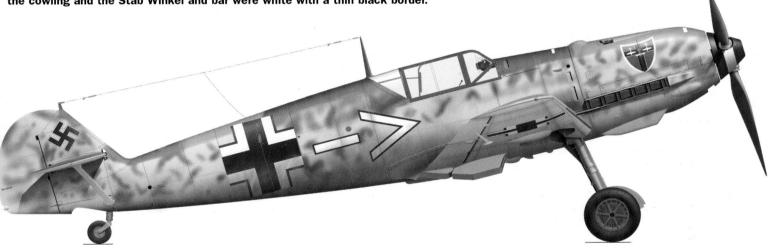

LEFT: The Bf 109 E-4 of Oblt. Albrecht Drehs of III./JG 54 photographed during the evening of 12 August as it lay in the field at Hengrove near Margate, Kent where it had force-landed after suffering damage from RAF fighters. The chevron and bar are white with a thin black outline and the band around the centre of the spinner is also believed to be white. The III.Gruppe emblem on the cowling consists of a red shield with a thin white border on which is superimposed a black and white 'Jesau' cross and three white aircraft silhouettes. As can be seen in the photograph, even at this early date, JG 54 had already begun to experiment with various patterns of additional colours to tone down the highly visible blue sides of their aircraft.

June-August 1940

**Devil emblem of
2./JG 52 (early)**

Messerschmitt Bf 109 E-1 of 2./JG 52 belonging to Uffz. Leo Zaunbrecher

'Red 14', the Bf 109 E-1 of Uffz. Leo Zaunbrecher of 2./JG 52 was forced to land in a field near Lewes after being damaged in combat by P/O J. McLintock of 615 Sqn. during the early afternoon of 12 August 1940. It carried a high demarcation upper camouflage of RLM 02/71 and the red painted 2.Staffel 'Little Devil' emblem superimposed on a white disc was painted on the port cowling only. The aircraft was also fitted with a rear view mirror mounted on the windscreen framing. Although shown in this profile with a red forward section, it is understood that some eye-witness reports stated that the spinner was red. However, whether these reports referred to the entire spinner or just the tip is not known for certain.

BELOW: August 12, 1940 and Uffz. Zaunbrecher's Bf 109 E-1 'Red 14' of 2./JG 52 lies in a cornfield at Mays Farm, near Selmeston, Sussex after being damaged in an earlier dogfight above Hastings.

RIGHT: Two policemen (possibly War Reserve Police Officers) take a close look at the cockpit of Uffz. Zaunbrecher's Bf 109 E-1 on the Sussex farm where it force-landed on 12 August 1940. Clearly seen in this view is the rear-view mirror attached to the top of the windscreen framework.

LEFT: RAF personnel gather around the nose of Uffz. Zaunbrecher's Bf 109 E-1 'Red 14' while one of their number attempts to remove the red devil emblem.

June-August 1940

ABOVE: Oblt. Wolfgang Ewald, Staffelkapitän of 2./JG 52, in discussion with airfield personnel at Calais in August 1940 following his return from a sortie. His Bf 109 E-3 'Black 1' can be seen behind.

LEFT: This salvaged Bf 109 E-1, 'Yellow 7' of 6./JG 51 has been transported to the airfield at Buc near Versailles prior to its being sent for repair. Although undated, it is believed that this photograph was taken during early August 1940.

Messerschmitt Bf 109 E-1 of 6./JG 51

Bf 109 E 'Yellow 7' of 6./JG 51 seen at the Aeroparc at Buc near Versailles during the summer of 1940. Finished in what is believed to be an upper surface scheme of 02/71, the fuselage sides are heavily mottled in one or both of these colours. The 6.Staffel weeping bird emblem aft of the Balkenkreuz carries a red umbrella with brown or black details while the body of the bird is understood to be brown with black and white details superimposed on a white shield with a thin black outline. It is not known with any certainty if the 'Gott Strafe England' motto was carried beneath the bird.

LEFT: The wreckage of one of the 38 RAF Blenheims which attacked the City of Aalborg (II./JG 77 area) on 13 August 1940 is towed away by a German truck for scrapping. The British force lost 15 of its aircraft during the raid.

RIGHT: Pilots of II./JG 77 at Aalborg in Denmark relax in the sun outside their dispersal hut which has been adorned with trophies obtained from the wreckage of British aircraft. II./JG 77 under Major Karl Hentschel, protected the northern reaches around Denmark, the Baltic and the Skagerak.

LEFT: This Bf 109 E-3 belonged to the Technical Officer of III./JG 51 is being refuelled at a temporary landing strip in France shortly after Dunkirk.

June-August 1940

ABOVE AND LEFT: Two views of the 3./JG 54 'Huntsman' emblem carried by some of that Staffel's aircraft during 1940. The cartoon huntsman figure, with a musket over one shoulder and carrying a brace of Spitfires was derived from the children's book 'Struwwelpeter'.

LEFT: Pilots of 1./JG 51 probably photographed at Pihen, France in early August 1940. From left to right: Ofw. Helmut Maul, Fw. Heinz Bär, Uffz. Herbert Biermann, Fw. Josef Oglodeck, Oblt. Hermann-Friedrich Joppien (Staffelkapitän from 6 August), Uffz. Thein, Uffz. Erwin Fleig and Uffz. Müller. Note the Mickey Mouse Staffel emblem at the entrance to the building. Of these pilots, Oglodeck was killed in action on 24 August. Bär would receive the Ritterkreuz (RK) with Eichenlaub (EL) and Schwerte; Joppien, the RK with EL and Fleig, the RK.

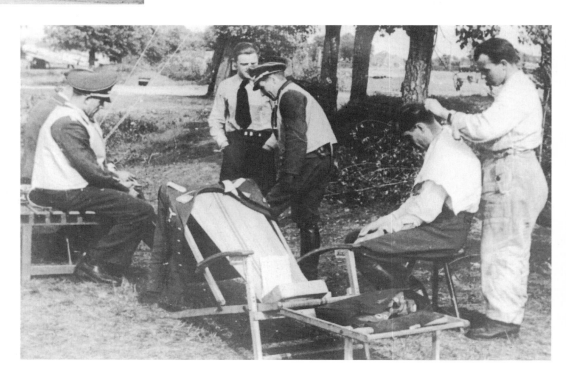

RIGHT: Members of I.(J)/LG 2 'relaxing' at readiness. Third from left is Herbert Ihlefeld.

RIGHT: 'Black 10', a Bf 109 E-1 of an unidentified Staffel, lies damaged in a French field. Interestingly, it has been fitted with the later style canopy but without the usually seen head armour. Note also this aircraft still has the gun ports edged in yellow.

LEFT: Assisted by his groundcrew, Oblt. Gustav 'Micky' Sprick, Staffelkapitän of 8./JG 26, straps in to his Bf 109 E-3 'Black 13' prior to a mission from Calais in mid-August. Note that this aircraft does not carry the Staffel 'Adamson' cartoon emblem.

BELOW: An unusual shot of four Bf 109 E-3s of 3./JG 3 with their engines running are preparing for a sortie. This view shows to advantage the variation of the splinter patterns on the uppersurfaces of the wings.

Messerschmitt Bf 109 E Camouflage and Markings 1939–1940

Confirming the identities of the camouflage colours and patterns worn by Bf 109 Es during the Battle of Britain presents an intriguing but complicated challenge. While it is known for the most part that the undersurface colour was usually a readily identifiable light blue, *e.g.* Light Blue 65 (RLM 65 *Hellblau*), the diversity in upper surface patterns and colours is far more difficult to ascertain.

A clear photograph of the nose and cowling of a Bf 109 E-1 showing the original low demarcation colour of RLM 70/71 over 65 camouflage finish.

Splinter Scheme or Single Colour?

A careful study of photographs of early Bf 109s reveals that the upper surface splinter camouflage patterns of Black-Green 70 (RLM *70 Schwarzgrün*) and Dark Green 71 (RLM *71 Dunkelgrün*) were applied with sharply defined, angular demarcation lines in keeping with standard *Luftwaffe* camouflage practice. The splinter patterns applied to the Bf 109B, C and D variants were similarly typical for the E-1 and E-3 which, as with the earlier models, displayed considerable variation on the fuselage sides where the pattern in plan view was extended down to meet the undersurface colour. This remained essentially unchanged until the final months of 1939 when a more simplified form of 70/71 splinter pattern began to make its appearance on some E models.

By the outbreak of war in September 1939, the camouflaged upper surfaces of Bf 109s were regularly identified as being 'dark green', implying the use of a single colour rather than the two dark greens officially specified by the *Reichsluftfahrtministerium (*RLM*)* in L..Dv.521/1 issued in March 1938. Did these observations accurately record that a single upper camouflage colour was being used or did the low tonal contrast between them prevent clear identification of

ABOVE: A clear photograph of two Bf 109 E-1s of I./JG 1 (formerly I./JG 131) which appear to be finished in a single upper surface colour. No evidence of a splinter pattern can be determined. Considering the wide variety of camouflage patterns applied to the Bf 109 and the fact that there were many different factories and sub-contractors manufacturing the type it is not beyond the possibility that one or two batches were produced in the one overall upper camouflage colour. It is also unlikely that completion would have been stopped because of a shortage of a specific camouflage colour.

III./JG 27
'Jesau Kreuz' emblem

Messerschmitt Bf 109 E-1 of I./JG 1, later III./JG 27

A Bf 109 E-1 'Red 9' of III./JG 27 which, from careful study of photographs, appears to have been finished in a single upper surface colour of either 70 or 71. As with a number of similar first generation Bf 109 E photographs that have been subjected to close scrutiny to date, no indication of a second upper surface colour has yet been determined. This has led to the belief that, for whatever purpose, some of these early models carried, at least temporarily, a single colour upper surface finish.

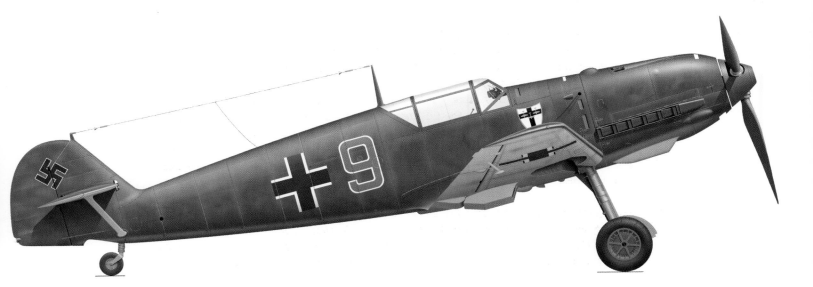

LEFT: The original print of the photograph above, and other views of this Bf 109 E-3 of II./JG 54 (I./JG 76) have been subjected to careful examination but no evidence of a second upper camouflage colour has been identified with certainty.

the two colours or, more simply, was this due to fading through in-service use and weathering?

During late 1939 – early 1940 and with the *Luftwaffe* fully committed to its wartime operations, the probability of a single upper camouflage colour being applied to individual aircraft or those of a specific unit is entirely credible. Although no valid or supportable documentary evidence of any Bf 109s with a single upper camouflage colour during this period has yet been discovered, it remains entirely plausible to assume that, for whatever reason, some aircraft *may* have received a single colour finish to the upper surfaces on either a temporary or permanent basis.

In the recent careful examination of a number of good quality original photographs, the presence of a single upper surface colour on some aircraft is strongly indicated as may be seen in the accompanying photographs. In the careful scrutiny of these original prints, to date, no discernible evidence of a second colour has been determined with complete certainty. Nevertheless, and until factual evidence to the contrary is discovered, it may perhaps be presumed that contemporary references to a single dark green are nothing more than a broad generalisation of the camouflage colour, the singular 'dark green' reference possibly being due to the low tonal contrast between these two colours?

RIGHT: The Bf 109 E-3 of Oblt. Hubert Kroeck of JG 53 as seen in November 1939. The aircraft was finished in the then standard low-demarcation camouflage of 70/71 over 65 undersurfaces.

1939-1940

The Colours Change

The fighting in Poland made it clear that while the standard 70/71 Bf 109 camouflage scheme at the outbreak of war was more than adequate for ground concealment, the same did not apply to aerial combat. As a result of these findings, numerous field trials to find a suitable replacement were undertaken during the winter of 1939-40 utilising various combinations of the colours *Grüngrau* (*aka* RLM) 02, 70 and 71. The successful outcome of these trials resulted in a new camouflage pattern of 02 and 71 that was considered more practical for air-to-air combat than the earlier scheme. Accordingly, an order was issued dictating that 02 would replace Black-Green 70 in the pattern. Concurrently, the demarcation for the undersurface Blue 65 was increased in height to cover approximately three-quarters of the fuselage sides, including the entire vertical tail surfaces. Although this change effectively restricted the upper colours to the strict plan view of the aircraft, the actual height of the demarcation varied considerably between aircraft, most prominently on the rear fuselage between the rear of the cockpit canopy and base of the fin.

Beginning in early 1940 with production of the Bf 109 E-4, the 02/71 scheme was applied as a factory finish, whereas the earlier E models already in service appear generally to have been repainted at either local or unit level, with the attendant broad interpretation of the contents of the order. Some units were noticeably slower than others in implementing the change and even for those repainting their aircraft promptly, it must be realised that not all aircraft within a single unit would or could be repainted at the same time. On some aircraft the 02 replaced the Dark-Green 71 rather than the Black-Green 70 as directed, while on many others, only the smaller areas of tail and fuselage were repainted, leaving the wing and fuselage upper surfaces in the earlier colours. That this occurred is evident from photographs and the contents of intelligence summaries, which indicates that a number of Bf 109s in these 'unofficial' finishes survived well into the early autumn of 1940. Furthermore, it is entirely possible that many of these may have been either older aircraft or those held as reserve or 'spare' aircraft, retaining their finishes until

BELOW: Newly-finished Bf 109 E-3s and E-4s, bearing the 'standard' high demarcation 71/02/65 finish await delivery to front line units. All of the Bf 109s seen here were scheduled for delivery to JG 3.

1939-1940

Two early 1940 photographs of 'Red 9' of 2./JG 2 in the newly-applied 71/02 scheme with the high-demarcation line for the underside RLM 65. Also visible on the upper surfaces of the wings is one of the many variations in the splinter camouflage pattern.

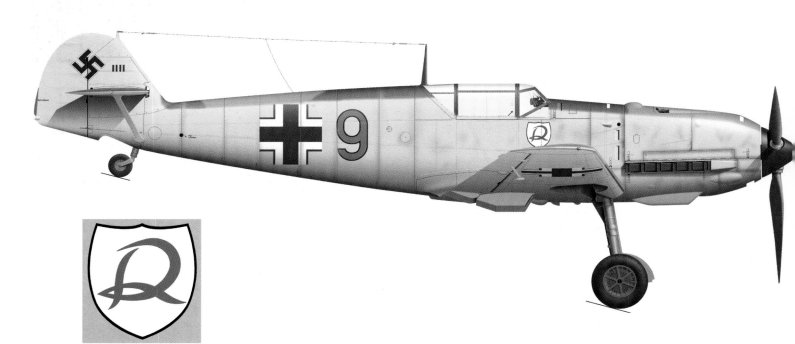

JG 2 'Richthofen'
Geschwader badge as
designed by
Lt. Adalbert von Rothkirch und Panthen

Messerschmitt Bf 109 E-1 W.Nr. 4859 'Red 9' of 2./JG 2
The aircraft is depicted prior to the beginning of the Battle of Britain when 2.Staffel was still using red numerals. Although the Hakenkreuz remains in the earlier position across both fin and rudder, it is finished in the high demarcation 02/71 scheme.

they were either lost on operations or underwent major servicing, at which time the newer scheme would, presumably, have been applied.

With these changes, which included revisions to the size, style and placement of the national insignia, several different examples of a simplified splinter scheme, including 'mirror' image reversal patterns, began to appear. In these, the colour divisions were far less angular than those of the original patterns and are often seen in photographs to have a 'feathered' rather than sharp demarcation. Although official confirmation for this simplification is unavailable, it is reasonable to assume that they were implemented as a means both to expedite service requirements and to save on materials and cost, regardless of whether the finish was of factory or in-service origin.

Summer 1940

As the aerial battles developed above the south-eastern coast of Britain and the English Channel in the early summer of 1940, it soon became clear that again, more changes would be necessary to the camouflage worn by Bf 109s. Whereas the 02/71/65 scheme had worked sufficiently well over France and the Low Countries, it was found that this was not the case in the air war against England. The high demarcation level of Blue 65 on fuselage sides and tails made the aircraft stand out conspicuously against the waters of the Channel and the English countryside. To overcome this, several methods were employed to tone down the blue, the most common being an application of mottling to fuselage sides in either 02 and/or one or both of the upper colours. One of the earliest reports of this occurred in mid-July when Bf 109s of JG 51 were reported as having a fine, pale grey 'overspray' applied

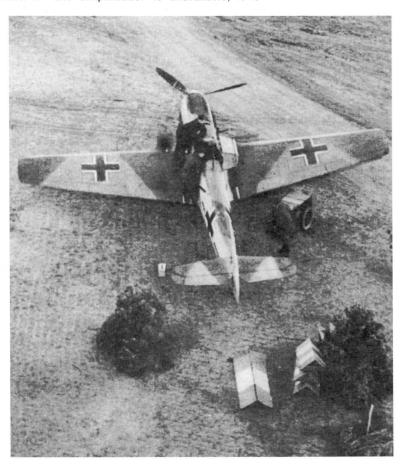

A clear overhead view of a Bf 109 E-1 which shows to advantage one of the simplified 02/71 splinter schemes that came into use during early 1940. This machine carries a single chevron and horizontal bar in front of the fuselage cross as well as a II. Gruppe bar behind.

to their fuselage sides; an indication perhaps of one of the first uses of 02 in this manner. Taking into account the requirement to tone these areas down, it is entirely feasible that an order was originated, either at RLM level or from local area command with *RLM* approval, allowing individual units, notably JG's 2, 53 and 54, to determine the extent and style of application as was dictated by their operational requirements. As the variations in mottling are far too extensive to describe in detail, it must be realised that while little similarity existed between individual units, a general uniformity of style and pattern was usually seen amongst aircraft of the same unit. Believed for the most part to have been 02, it was usually sprayed on the sides of the fuselage and fin in varying degrees of density and pattern. On some aircraft this was occasionally intensified, usually where a colour transition was made such as at the roots of the wings or tailplane, by the random inclusion of one or both of the upper colours. By contrast, the mottle applied by some units was in a much coarser form, suggesting the use of a brush or sponge, frequently so dense that it took on the appearance of an almost solid colour. Noticeably, many aircraft wearing this coarse, stippled finish (e.g. JG 2) also displayed a modified fuselage cross where the proportions of the white segments were reduced in area to decrease their visibility. Similarly, the height of demarcation between upper and lower colours was often altered, with segments of the upper fuselage colours being extended down the fuselage sides to random depths along its length. However some units, notably the third *Gruppe* of JG 26, were markedly reluctant to add any form of additional camouflage to their aircraft and, throughout 1940, retained the high demarcation finish with fuselage crosses and numerals repainted in a smaller format than usual to help conceal the aircraft at higher altitudes.

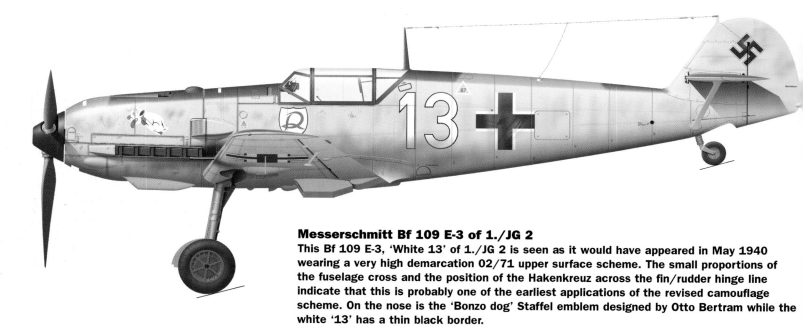

Messerschmitt Bf 109 E-3 of 1./JG 2
This Bf 109 E-3, 'White 13' of 1./JG 2 is seen as it would have appeared in May 1940 wearing a very high demarcation 02/71 upper surface scheme. The small proportions of the fuselage cross and the position of the Hakenkreuz across the fin/rudder hinge line indicate that this is probably one of the earliest applications of the revised camouflage scheme. On the nose is the 'Bonzo dog' Staffel emblem designed by Otto Bertram while the white '13' has a thin black border.

**1./JG 2
'Bonzo dog'
Staffel emblem**

With more fighter engagements taking place over the sea and increasing numbers of replacement aircraft entering service, camouflage variations became all the more widespread, often becoming more varied when easily interchangeable parts such as cowlings, rudders, armament access panels and battery hatch covers were swapped between aircraft to expedite servicing. Additionally, two further anomalies appeared for which, to date, no firm explanation has been determined. The first was a lighter centre to wing crosses that may or may not have been a part of the random light camouflage overspray occasionally seen on wings or was perhaps, evidence of the overall mottled finish seen and documented as being applied to some Bf 109 Es during 1940. Although no documentary evidence to support this has been found, it is reasonable to assume that the additional colour was applied to reduce the visibility of the wing crosses and blend them in to the upper surface camouflage, thus helping to conceal the aircraft from observation from above.

The second anomaly and one which is evident in many photos of Bf 109s from the period, was the use of a light colour that wrapped around the upper leading edges of the main wings and may clearly be seen in photos such as those of the aircraft of *Oblt.* Paul Temme of *Stab*/JG 2 who force-landed beside Shoreham airfield on 13 August. From the detailed examination of photographs of aircraft with this feature it is currently believed that this was in fact, a continuation of the underside Blue 65 or similar light colour, extended to encompass the areas of wing leading edge visible in a head-on view. Whether this was an attempt to break up the outline of the wings when viewed from head-on or an application characteristic of the location where the camouflage finish was applied has not, to date, been determined with any certainty.

ABOVE: A formation of Bf 109 E-3s of 1./JG 2 on patrol in May 1940. The 'Bonzo dog' Staffel emblem illustrated to the right can be seen on the cowling of 'White 7'.

RIGHT AND BELOW: An example of one of the adaptations of fuselage mottling applied by JG 54 can be seen in this view of 'Yellow 13' of 9. Staffel. Also clearly visible is the black bordered yellow Staffel shield containing the devil's head emblem; the head is red with black and white details.

Messerschmitt Bf 109 E-4 of 9./JG 54

The photograph above and this colour profile of 'Yellow 13', a Bf 109 E-4 of 9./JG 54, illustrate one of the variations in the striped mottling commonly used by this unit. In this instance, diagonally sprayed stripes of what appears to be 71 have been applied to break up the blue of the fin, rudder and fuselage sides.

**9./JG 54
'Devil's Head' emblem**

RIGHT: Bf 109Es of II./JG 2 take off on a sortie from an unidentified airfield in France. While the first aircraft appears to be finished in the basic 71/02/65 scheme, the second – Black or Red '4' – has heavily mottled sides with the mottling also applied to the tops of the wings and tail planes. It also carries the Hakenkreuz in the early position across the fin/rudder hinge line and early type Balkenkreuz.

ABOVE: Set up for weapons calibration, this Bf 109 E of an unidentified unit illustrates another of the variations of mottling applied over the blue 65 of the rudder, fin and fuselage sides.

ABOVE: The Bf 109 E-3 'White 2' flown by Hans Illner of 4./JG 51. The photograph portrays how the visibility of the upper wing Balkenkreuze have been subdued by a light overspray of paint. Also evident in this view is the light random application of RLM 02 patches to the upper surfaces of wings and tailplanes.

Grey Camouflage?

Although often totally destroyed, all enemy aircraft that came down in the British Isles during the Second World War were thoroughly examined by intelligence teams from the Air Ministry and RAF. The reports created from these examinations were known as Crashed Enemy Aircraft Reports, and recorded such information as *Werk Nummer*, engine type, armament, additional or special equipment and markings and colours. However, and to the disappointment of many post-war researchers, there were no set guidelines in these reports for describing the shades of the colours found on downed German aircraft. Generally, any examination of the paint was confined to an evaluation of the type of finish and occasionally, some undamaged panels would be tested for paint durability.

By mid-August, the first uses of greys and blue-greys as an upper camouflage colour were making their appearance in these reports, appearing with increasing frequency as the battle progressed. 'Light Navy grey', 'two shades of grey', 'light grey with dark grey mottling', 'Battleship grey' and 'camouflage grey' were some of the descriptions recorded, along with mention of varying shades of green-grey and blue-grey. Were these an indication of the earliest use of the greys 74 (RLM *74 Dunklegrau*) and 75 (RLM *75 Mittelgrau*) that would become the standard fighter camouflage the following year or, as recent research and the variety in their descriptions suggest, that they were colours created at unit level?

As illustrated on page 89, it can be seen that by mixing various percentages or combinations of RLM 02, 65, 66, 70 and 71, a number of grey and blue-grey shades could have been created, all of which would have been suitable for use, thus providing a perfectly credible probability that this is, in fact, what happened. As the use of the greys 74 and 75 was not officially promulgated until the November 1941 issue of L.Dv 521/1, the likelihood that the assorted greys used during 1940 were those from which the 74 and 75 were developed is a wholly convincing possibility.

LEFT: As may be seen here, the Balkenkreuze on the upper wing surfaces of a number of Bf 109s during mid to late 1940 were often partially covered by light mottling to subdue their visibility. This is the upper wing of Lt. Johann Böhm's Bf 109 as featured on page 24.

JG 26 'Schlageter' emblem

Messerschmitt Bf 109 E-1 of 6./JG 26 May 1940
Seen at Brugelette, Belgium at the end of May 1940, 'Yellow 2' of 6./JG 26 was already wearing a lightly applied mottle to the upper surface 02 and 71 on the fuselage sides while beneath the windscreen can be seen the stylised JG 26 'Schlageter' shield emblem.

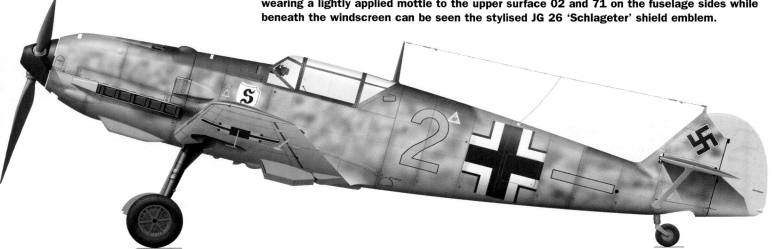

LEFT: Brugelette, 31 May 1940 and ground crews of II./JG 26 sit in the late spring sun to service their personal weapons. Of the aircraft seen in this photograph, 'Yellow 2' already has a light mottle applied over its RLM 65 sides while the aircraft in the background does not. 'Yellow 2's' II. Gruppe bar is hidden by the mechanic sitting on the right.

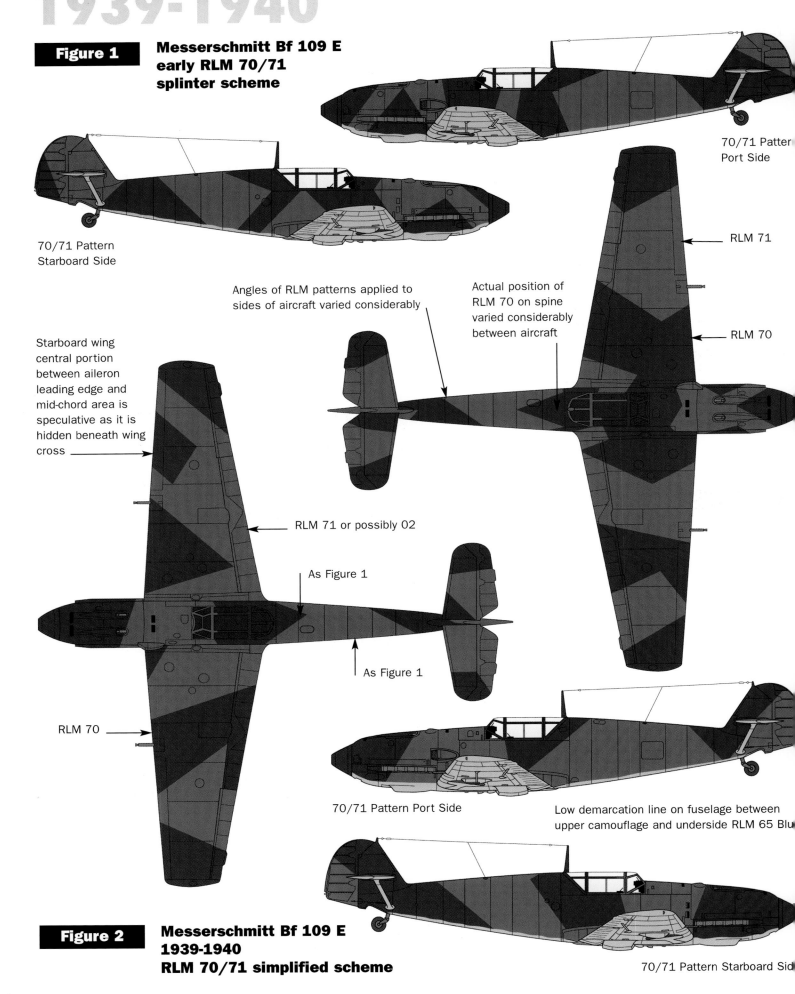

Figure 1

**Messerschmitt Bf 109 E
early RLM 70/71
splinter scheme**

70/71 Pattern
Port Side

70/71 Pattern
Starboard Side

RLM 71

RLM 70

Angles of RLM patterns applied to
sides of aircraft varied considerably

Actual position of
RLM 70 on spine
varied considerably
between aircraft

Starboard wing
central portion
between aileron
leading edge and
mid-chord area is
speculative as it is
hidden beneath wing
cross

RLM 71 or possibly 02

As Figure 1

As Figure 1

RLM 70

70/71 Pattern Port Side

Low demarcation line on fuselage between
upper camouflage and underside RLM 65 Blu

Figure 2

**Messerschmitt Bf 109 E
1939-1940
RLM 70/71 simplified scheme**

70/71 Pattern Starboard Sid

Demarcation height variations

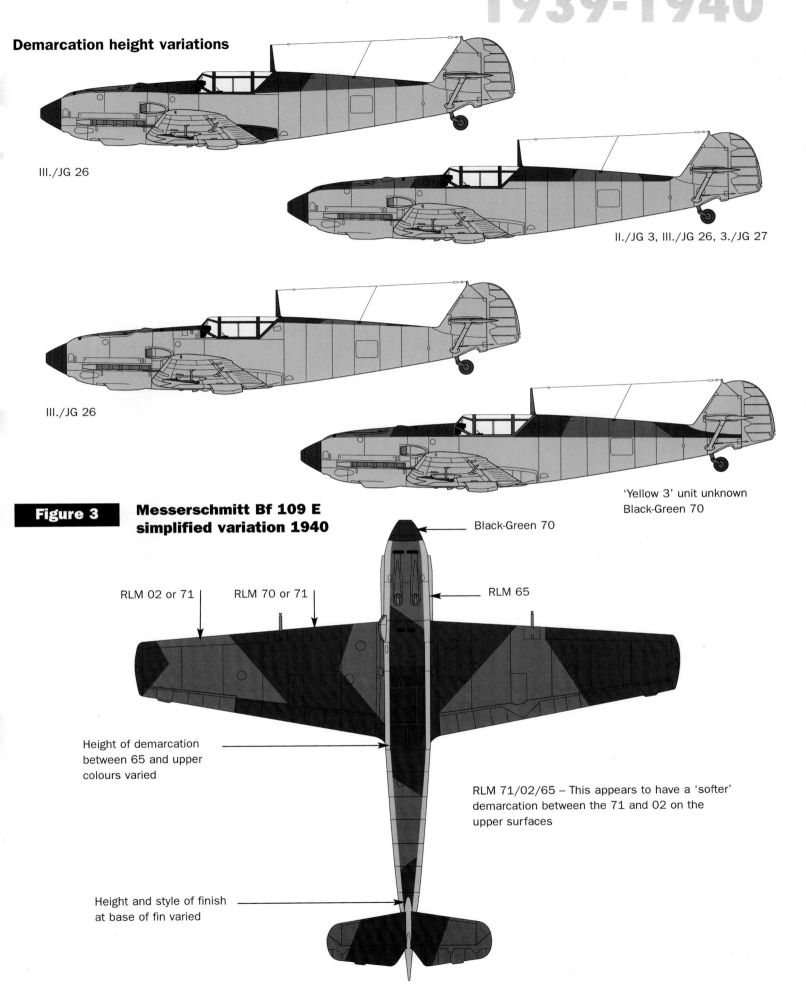

III./JG 26

II./JG 3, III./JG 26, 3./JG 27

III./JG 26

'Yellow 3' unit unknown
Black-Green 70

Figure 3 **Messerschmitt Bf 109 E simplified variation 1940**

Black-Green 70

RLM 02 or 71

RLM 70 or 71

RLM 65

Height of demarcation between 65 and upper colours varied

RLM 71/02/65 – This appears to have a 'softer' demarcation between the 71 and 02 on the upper surfaces

Height and style of finish at base of fin varied

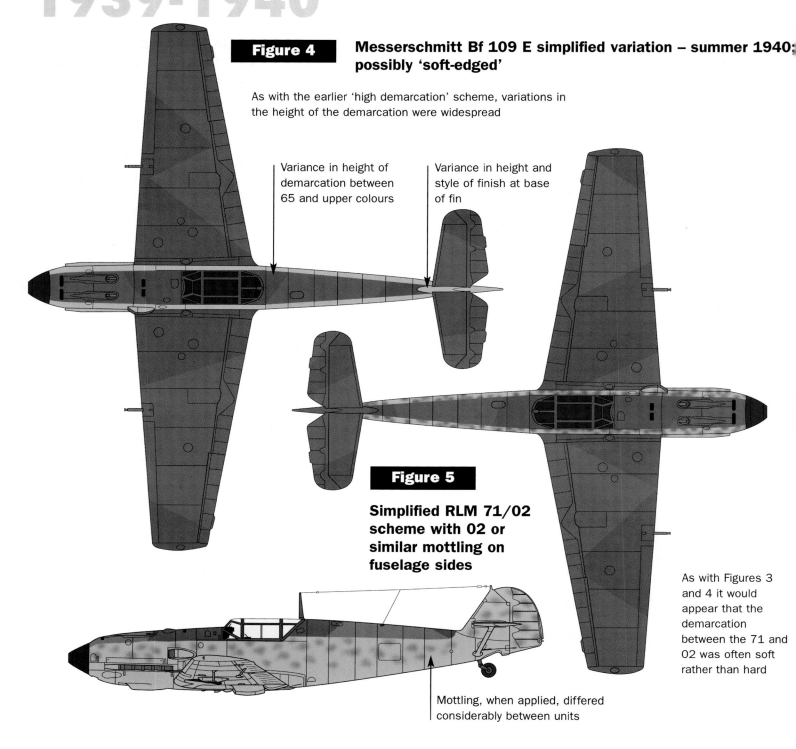

Figure 4

Messerschmitt Bf 109 E simplified variation – summer 1940; possibly 'soft-edged'

As with the earlier 'high demarcation' scheme, variations in the height of the demarcation were widespread

Variance in height of demarcation between 65 and upper colours

Variance in height and style of finish at base of fin

Figure 5

Simplified RLM 71/02 scheme with 02 or similar mottling on fuselage sides

As with Figures 3 and 4 it would appear that the demarcation between the 71 and 02 was often soft rather than hard

Mottling, when applied, differed considerably between units

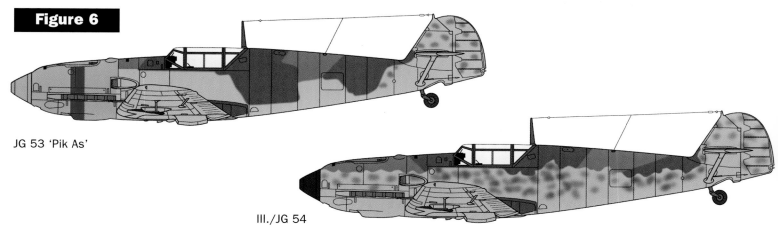

Figure 6

JG 53 'Pik As'

III./JG 54

Four examples of Grey shade variations made from various RLM paint mixes

Although difficult to represent accurately in printed form, these four panels serve to illustrate an approximation of some of the grey shades that could be obtained by combining various quantities of readily available paints such as 02, 65, 66, 70 and 71.

RLM 65/66

RLM 02/65/66

RLM 66/02

RLM 70/65

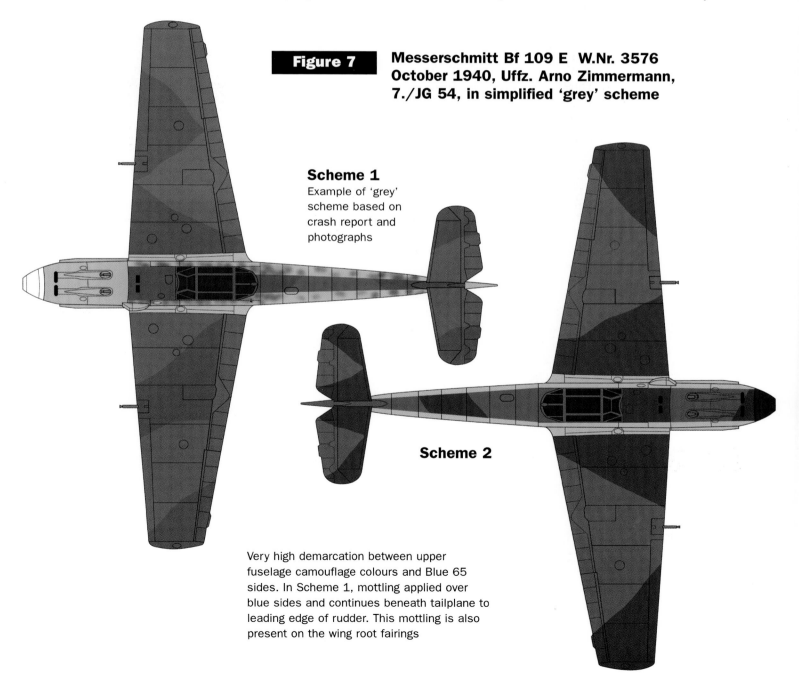

Figure 7

Messerschmitt Bf 109 E W.Nr. 3576 October 1940, Uffz. Arno Zimmermann, 7./JG 54, in simplified 'grey' scheme

Scheme 1
Example of 'grey' scheme based on crash report and photographs

Scheme 2

Very high demarcation between upper fuselage camouflage colours and Blue 65 sides. In Scheme 1, mottling applied over blue sides and continues beneath tailplane to leading edge of rudder. This mottling is also present on the wing root fairings

Gruppenstab and Staffel Markings and Colours

Stab Markings

The origins of markings for *Stab* personnel date back to the pre-war period, being allocated to three officers of the *Geschwaderstab:* the *Kommodore*, his *Adjutant* and the officer in charge of flying operations. Correspondingly for the *Gruppenstab*, similar symbols were allocated to the *Gruppenkommodore*, *Adjutant* and Operations officer.

The entry into service of the Bf 109 made it apparent that the earlier *Stab* symbols would need revising. Accordingly, *Fl.Inst. 3 Nr.730/37 II 9* issued on 14 December 1937 by the *Generalstab der Luftwaffe* included a set of instructions and diagrams for the application of markings to fighter aircraft. Apart from containing detailed instructions on the dimensions of numerals and their spacing, new locations and dimensions for *Stab* symbols were designated, including a vertical bar symbol to be applied aft of the fuselage cross to signify III.*Gruppe* instead of the earlier wavy line symbol. A horizontal bar aft of the fuselage cross identified the aircraft of II.*Gruppe* while those from I.*Gruppe* carried no symbol. All symbols were to be applied in black with white edging and a thin black outline although it is evident in photographs and other records that these markings were not always applied in either the colours or locations officially specified. Despite the clearly worded instructions regarding the III.*Gruppe* marking, at least two *Jagdgeschwader*, JG 2 & JG 52 declined to effect the change, retaining instead the earlier wavy line *Gruppe* symbol. In similar fashion, III./JG 2 and JG 54 also declined to follow the wording of the directive by using white as the predominant colour for their *Stab* symbols, usually outlining them with a thin black edge. Likewise, II./JG 51 also declined to display their *Gruppe* bar by using the designated area to display their 'weeping bird' emblem instead of the required symbol.

Staffel Markings

The ordinary *Staffel* aircraft carried a number which identified the individual aircraft within the *Staffel*, and the colour in which it was painted identifying the *Staffel* within the *Geschwader*. These numbers were generally applied in one of two forms with the figures from 2 to 9 appearing in either a 'rounded' or 'squared' style that usually remained constant within the various *Staffeln*.

Although regularly positioned ahead of the fuselage cross, some units did adopt alternative locations for these numbers. III./JG 27 chose to apply them to either side of the cowling beneath the gun troughs while III./JG 54 placed theirs on either side of the forward fuselage, just aft of the rear edge of the engine cowling. Likewise, there were also exceptions to the rule for *Staffel* colours; on several occasions, red was recorded as replacing the normal black of the second, fifth and eighth *Staffeln*, the third (*Jagd*) *Staffel* of LG 2, used brown instead of yellow and 5.*Staffel* of JG 53 is recorded as using grey numerals throughout 1940.

Spinners too received their share of colours. These were often repainted in black and white in the form of halves or quarters or would merely have a segment of white applied to the base Black-Green 70 spinner colour. In many instances the spinner tip or cap, if fitted, would often be painted in the *Stab* or *Staffel* colour. While there are no reports of the 1944 '*Spiralschnause*' style of design being used at this time, those coloured bands that were painted on Bf 109 E spinners during 1940 are recorded as being applied in *concentric* circles.

Colours

I, II & III *Gruppenstab*	Green
1, 4 & 7 *Staffeln*	White
2, 5 & 8 *Staffeln*	Black (or Red)
3, 6 & 9 *Staffeln*	Yellow (or Brown)

The Red Band of JG 53

For a short period during 1940, all three *Gruppen* of JG 53, and only JG 53, displayed two distinct anomalies in their markings, the purposes of which have yet to be fully resolved.

The first anomoly concerns the replacement of the 'Pik As' (Ace of Spades) emblem. According to RAF Air Ministry Weekly Intelligence Summary No.60, Hermann Göring ordered the emblem be removed and replaced with a red band and the Geschwader renamed the 'Red Ring

LEFT: La Villiaze on the island of Guernsey was used as a forward base by JG 53 for its cross-Channel operations. In this view of Lt. Gerhard Michalski posing beside the aircraft of Hptm. Günther von Maltzahn, the red ring carried by the unit's Bf 109s can clearly be seen.

Messerschmitt Bf 109 E-3 of II./JG 53

Identified as the Bf 109 E-3 belonging to Hptm. Freiherr Günther von Maltzahn, Gruppenkommandeur of II./JG 53, this profile illustrates one of the variations of the red cowling band carried by aircraft of the unit between August and October and which replaced the unit's Pik-As emblem. In early October, shortly after von Maltzahn was promoted to take over command of the Geschwader from Major von Cramon-Taubadel, the red band was dispensed with and replaced with a newer style of Pik-As emblem.

Geschwader. While there is some evidence to suggest that it may have stemmed from some personal antipathy on the part of Göring, or possibly from some ideological difference with the leadership of the *Geschwader*, (H-J von Cramon-Taubadel is understood to have had a Jewish wife), the actual reason for the order has yet to be determined. In the past, several valid theories for this change have been examined in depth, but most have been subsequently disproven although one, containing some merit, submits that it may have been nothing more than a temporary identification feature. However, there was one event which transpired at this time and another which may have been of some significance. During early August, at around the time of the appearance of these red bands, Göring replaced the majority of the *Jagdwaffe Kommodore* with younger men, although two units serving with *Luftflotte* 3, JG 27 and JG 53, retained their existing *Kommodore* until October. Then, at the beginning of that month, after *Oblt*. Günther von Maltzahn took command of the *Geschwader* from *Oblt*. Hans-Jürgen von Cramon-Taubadel, the 'Pik As' emblem began to reappear on JG 53's aircraft in a somewhat newer and larger format than previously seen. As a matter of interest, the first recorded incident of a Bf 109 E being brought down over England where the red band had replaced the 'Pik As' emblem occurred on 16 August. On that date the aircraft of *Fw.* Christian Hansen of 2./JG 53 force landed at Godshill on the Isle of Wight and when examined was reported in Crashed Enemy Aircraft Report No.11 as having a "...red band around nose 6 in wide".

The second anomaly, and one frequently recorded as a political gesture on the part of the *Geschwader,* occurred almost concurrently with the reintroduction of the 'Pik As' emblem. Many aircraft from II. and III *Gruppen* had the *Hakenkreuz* on their fins overpainted, with several pilots using these areas to display their individual *Abschuss* tallies rather than in the more usual location on the rudder (e.g. *Lt.* Schmidt, *Adjutant* of III./JG 53). How long this lasted is not known for certain but some aircraft of III. *Gruppe* were recorded as still without their *Hakenkreuz* in late November.

Messerschmitt Bf 109 E-4 of III./JG 53
The Bf 109 E-4 of Lt. Erich Schmidt, Adjutant of III./JG 53 in November 1940. Finished it what is understood to be an upper scheme of locally mixed greys, the 'Pik-As' emblem has replaced the 'Red Ring' and like other aircraft of the Gruppe, the Swastika has been painted out; in this case being replaced by the pilot's 'Abschuss' tally.

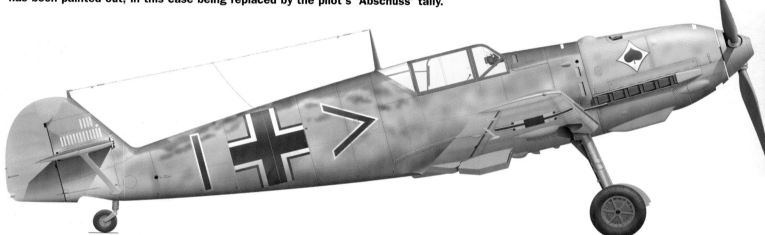

JG 53 Pik As Geschwader

BELOW: Photographed at Etaples during November 1940, these two aircraft of the Gruppenstab III./JG 53 illustrate the second markings anomaly seen on the aircraft of this unit – the overpainted Hakenkreuz. The aircraft in the foreground is that of the adjutant, Lt. Erich Schmidt, while that in the background with the 'Pik As' emblem clearly visible on the cowling is that flown by the Kommandeur, Wolf Wilcke.

Tactical Markings

The requirement that visually, a military aircraft should be invisible to its foe but instantly recognisable to friendly forces is something of a practical impossibility, and throughout the history of military aviation, numerous methods to resolve this problem have been examined. With the *Luftwaffe* it was no different. In mid-August, the first incidents involving Bf 109s carrying distinctive yellow markings were being reported by RAF pilots. Originally interpreted as denoting aircraft belonging to a 'squadron of aces', this assumption was incorrect.

The earliest examples of the use of these markings occurred when aircraft of JG 26 and JG 54 were recorded as carrying areas of yellow paint applied to wing and tailplane tips and also to top sections of rudders and on occasion, to the vertical trailing edge section of the rudder. There is little doubt that these markings were established as an aid to instant recognition in the air where such conspicuous markings were invaluable to both sides. In appreciation of this value, the *Jagdwaffe* were quick to increase the use of such colours to include cowlings and entire rudders. Whereas the application of either yellow or white paint to wing and tailplane tips remained relatively constant from unit to unit, this was often not the case where cowlings and rudders were concerned.

On rudders, it first appeared in the form of an inverted triangular area on the top section as may be seen in photographs of Gerhard Schöpfel's Bf 109 of III./JG 26 circa mid-August. Not long after this, other Bf 109 Es, often noted as being from III./JG 54, were recorded as having approximately one-third of the rear vertical rudder surface painted yellow or possibly, white, while on the Bf 109s of other units, the entire rudder was finished in one of these colours. When the whole rudder was painted, the exact area covered often varied as occasionally, a section of the original Blue 65 would be left on which the pilot would display his 'Abschuss' tally, usually marked as black or red vertical bars that often identified the nationality of the victim and the date of the victory. In addition to these variations, at least two Bf 109 Es of the period are documented where the entire fin and rudder were also painted in yellow but based on currently available information, these are seen to be the exception rather than the rule at this time.

LEFT: The larger format JG 53 'Pik As' as applied after the removal of the red ring. The aircraft in the photograph is also fitted with one of the two styles of nosecaps commonly seen on Bf 109 Es during the second half of 1940.

LEFT: Bf 109 E-3 'White 6' of 7./JG 26 on which the early application of yellow paint to the top portion of the rudder can clearly be seen.

7./JG 26 emblem

Messerschmitt Bf 109 E -3 of 7./JG 26

'White 6' of 7./JG 26 illustrates the early form of yellow rudder marking that was beginning to make its appearance on Bf 109s during the middle of August 1940. Initially starting off as shown here, it later increased to cover the rear half of the rudder surface and finally the entire surface.

With cowlings, it can be seen from photographs that the area covered by white or yellow paint varied considerably between aircraft, often extending rearwards as far as the base of the windshield. Any unit emblems that would otherwise be hidden by this paint were usually masked off carefully, and two such units, I./JG 3 and III./JG 27, masked off the distinctive JG 3 '*Tatzelwurm*' and JG 27 numbers so as to leave them on a conspicuous rectangular background of the camouflage colour. From late August on, it is unusual to find a photograph of a Bf 109 E without some part of its airframe covered in either yellow or white paint, and to date, no significant explanation for the use of the two different colours has been ascertained, suggesting that they may have been used somewhat indiscriminately. In addition to the use of

Messerschmitt Bf 109 E-1 of 9./JG 51

'Yellow 13' of 9./JG 51, finished in a high demarcation upper scheme of 02/71,and illustrates the early application of yellow paint to rudders which took the form of an inverted triangle at the top of the rudder. Beneath the lower edge of the canopy is carried the III.Gruppe emblem of the 'Axe of the Lower Rhine' which had originated with the Gruppe when it was 1./JG 20.

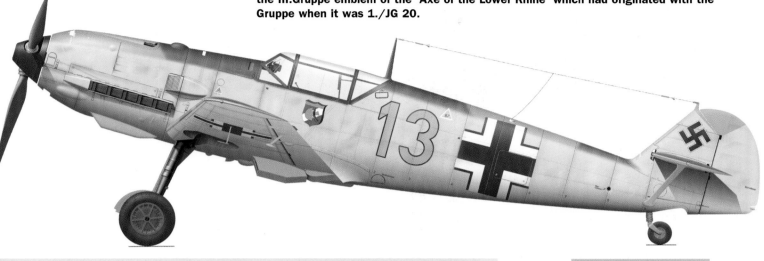

**9./JG 51
'Axt von Niederrhein
emblem**

ABOVE: Badly damaged, Bf 109 E-1, 'Yellow 13' of 9./JG 51 lies on a French beach during the summer of 1940. Seen to advantage is the early application of yellow paint to rudders which took the form of an inverted tri-angular area at the top of the rudder.

yellow and white for these tactical markings, it is also claimed by some sources that red was likewise used for the same purpose. However, despite several detailed investigations, no photographic or documentary evidence whatsoever has been discovered to support this.

Although some references suggest that the change from yellow to white occurred at the end of August, it is evident from the contents of Crashed Enemy Aircraft Reports for the month of September that both colours were being used concurrently by different units during that time. As far as current research has shown, it would appear that this use of white lasted only for a period of approximately three or four weeks and was seemingly confined in the main to units based within a small sector of occupied France. During the last week of August, the fighter units of *Luftflotte* 3 were placed under the control of *Luftflotte* 2 when the bomber units of the former were temporarily withdrawn from daylight operations in order to join the nightly attacks on centres of industry in the Midlands. However, whether or not this was in any way connected with the use of the white tactical markings for the single-engined fighter force, remains a matter of speculation for the present.

Summary

While it is a well-recognised fact that the *RLM* had a clearly defined administrative intent to regulate *Luftwaffe* camouflage practices, it must also be distinctly understood that, as surviving documentary and photographic evidence reveals, there were many exceptions to its established edicts. Unfortunately, since very few original documents or diagrams are available from which definitive information can be obtained, much of the interpretation for these variations must rely heavily on informed and educated speculation based upon such material and knowledge as is currently available.

BELOW: Bf 109 E-3s of 3(J)./LG 2 sit concealed beneath the trees at Calais-Marck while unit personnel improve the effect by suspending camouflage nets above them. In the foreground, 'Brown 11', in similar fashion to other aircraft of the unit, has only the rear half of the rudder painted yellow. The speckled appearance of its upper surfaces are not a mottled paint scheme but due to the sunlight filtering through the branches and camouflage netting. Not the unknown badge behind the fuselage cross.

ABOVE: The church spire in the distance identifies this as Calais-Marck airfield where the fighters of 3(J)./LG 2 are preparing for a mass take-off. On the rudder of the Bf 109 E-3 at left it can be seen that only the rear half carries yellow paint, a not uncommon practice on aircraft of this unit. Later in the war the church was bombed by the USAAF during a raid on the airfield.

Figure 8

Yellow and White Areas (approximation of areas covered when yellow or white used)

This photo taken during the later stages of the battle serves to illustrate where white or yellow paint has been applied to both upper and lower cowlings and spinner. In this view it can be seen that the backplate has been left in a darker colour, probably 70, and a cap has been fitted to the front of the spinner.

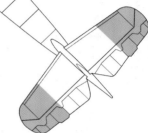

Covered area on spinner and cowling varied considerably

Schöpfel's (JG 26) and Tiedmann's (JG 3) Bf 109s carried yellow wing and tail tips in mid-August 1940 and both aircraft had yellow top segments to the rudders

Areas shown in yellow here were similarly painted white by some units; ie: Uffz. Wolff's Bf 109 E of 3./JG 52 and Oblt. von Werra's Bf 109 E of JG 3

Figure 9

Examples of variations in the application of yellow or white tactical(?) markings, Summer 1940

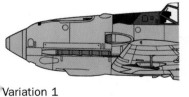

Variation 1
JG 3, JG 54

Note: Spinner colours varied considerably ranging from completely yellow or white or black-green to various permutations.

Von Werra, II./JG 3

Uffz. Wolff, 3./JG 52

X

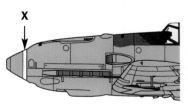

Note: Wick's Bf 109 with an all-yellow nose had a white backplate – marked here with 'X'

As per 'Assi' Hahn's Bf 109 E of I./JG 3

Soft demarcation between forward edge of yellow and remainder of camouflage paint on rudder – i.e. Zimmermann's Bf 109 E

Variation 2
JG 2 (Wick), JG 27, 8./JG 54

Yellow segment applied to top of rudder – August i.e. Schöpfel, JG 26 and Tiedman, JG 3

Yellow paint applied only to rear section of rudder as per some Bf 109 Es of 3(J)./LG 2.

Variation 3
ie: 'Assi' Hahn's Bf 109 E, I./JG 3

I./JG 3, von Werra II./JG 3, JG 26, 3./JG 52

RIGHT AND BELOW: A close-up view of the cockpit and spinner areas of a force-landed Bf 109 E-3 of JG 2 clearly showing the heavily applied green (RLM 70) brush or sponge stipple that was beginning to be applied by the ground crews of this unit at this time. The aircraft was damaged by machine gun bullets which penetrated the engine and cockpit side panels. The broken plexi-glass shows how thin and vulnerable the material was which led to an additional armoured glass panel being applied to the front windscreen on many aircraft. Note part of the JG 2 'Richthofen Geschwader' badge below the cockpit which has been crudely stippled around.

BELOW: Claimed to be the aircraft of Werner Machold, this rare colour photograph showing an 'Abschuss' tally of twenty-three victories, is believed to date from 28 or 29 August when III./JG 2 returned to Querqueville from Calais. On 30 August Machold would increase his score to twenty-four with the claim for another victory near Portland.